STUDIES IN GAME THEORY AND MATHEMATICAL ECONOMICS

ABOUT THIS SERIES

Game theory, since its creation in 1944 by John von Neumann and Oskar Morgenstern, has been applied to a wide variety of social phenomena by scholars in economics, political science, sociology, philosophy, and even biology. This series attempts to bring to the academic community a set of books dedicated to the belief that game theory can be a major tool in applied science. It publishes original monographs, textbooks and conference volumes which present work that is both of high technical quality and pertinent to the world we live in today.

EIGHTY-NINE EXERCISES WITH SOLUTIONS FROM GAME THEORY FOR THE SOCIAL SCIENCES, SECOND AND REVISED EDITION

Hervé Moulin

NEW YORK UNIVERSITY PRESS
NEW YORK AND LONDON
1986

Library of Congress Cataloging-in-Publication Data

Moulin, Hervé.
89 exercises with solutions from Game theory for
the social sciences, 2nd and revised edition.

(Studies in game theory and mathematical economics)
Includes index.
1. Social sciences—Mathematical models.   2. Game
theory.   I. Moulin, Hervé. Théorie jeux pour l'économie
et la politique. English. 2nd rev. ed.   II. Title.
III. Title: Eighty-nine exercises with solutions from
Game theory for the social sciences, 2nd and revised
edition.   IV. Series.
H61.25M678 1986   300'.1'5193   86-5315
ISBN 0-8147-5432-5
ISBN 0-8147-5433-3 (pbk.)

The medallion on the cover of this series was designed
by the French contemporary artist Georges Mathieu as one of
a set of medals struck by the Musée de la Monnaie in 1971.
Eighteen medals were created by Mathieu to "commemorate 18
stages in the development of western consciousness." The
Edict of Milan in 313 was the first, Game Theory, 1944, was
number seventeen.

INTRODUCTION

"The only way to learn mathematics is by doing exercises." The old saying plausibly extends to game theory, among others.

The second edition of my course "Game theory for the social sciences" proposes 89 exercises (many of them with the length and technical level of true problems) accumulated during nine years of teaching. A detailed solution of those will hopefully enhance their pedagogical added value. At any rate they will help the instructor.

We have reproduced the statement of the exercises (systematically embellished by a name) as they appear in the second edition of the textbook. Definitions and notations are not duplicated, and neither are the examples from which some of the exercises are derived.

H. Moulin

CONTENTS

CHAPTER 1   TWO-PERSON ZERO-SUM GAMES

a)   *Normal form games: examples*

1)   Gale's roulette

Each wheel has an equal probability to stop on any of its numbers. Player 1 chooses a wheel and spins it. While it is still spinning Player 2 chooses another wheel and

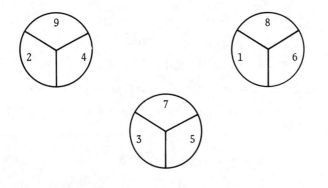

spins it. The winner is the player whose wheel stops on the bigger score. He receives $1 from the loser.

Which player would you like to be in this game?

If you are the winning player, at what price would you sell your seat in the game to a third party (assuming you are risk neutral and compare random events by their expected payoff)?

Solution:

The odds are 5 to 4 for A against B, and for B against C, and for C against A. Hence Player 2 secures an expected gain (5/9) - (4/9) = $1/9.

2) Land division game

The land consists of three contiguous pieces: a square with corner (0, 0), (1, 0), (1, 1), (0, 1); a triangle with corners (0, 1), (1, 1), (0, 2); and a triangle with corners (1, 0), (1, 1), (2, 1). The method of division is as follows. Simultaneously Player 1 chooses a vertical line L_v with the x coordinate in [0, 1], and Player 2 chooses a horizontal line L_h with the y coordinate in [0, 1]. Then Player 1 gets all the land which is above L_h and to the left of L_v as well as all the land which is below L_h and to the right of L_v. Player 2 gets the rest. Find the value of the game and optimal strategies, assuming both players want as much land as possible.

3

Solution:

The normal form of the game is

$$X = Y = [0, 1]$$

$$u(x, y) = 2\mathbf{x} + y - 2xy + \frac{1}{2}(y^2 - x^2)$$

$$v(x, y) = 2 - u(x, y)$$

Check u is concave in x, convex in y, so by von Neumann's theorem there is a saddle point. Seeking it as an interior point $0 < x^*, y^* < 1$ it is enough to solve:

$$0 = \frac{\partial u}{\partial x}(x^*, y^*) = 2 - 2y^* - x^*$$

$$0 = \frac{\partial u}{\partial y}(x^*, y^*) = 1 - 2x^* + y^*$$

So $x^* = (4/5)$, $y^* = (3/5)$, value: $u^* = (11/10)$, $v^* = (9/10)$.

3) <u>Silent gunfight</u>

This is a variant of Example 1. However, here a shot is fired which cannot be heard. Thus you are not aware that your opponent has shot unless you are hit. The payoff function is now:

$$u(x_1, x_2) = \begin{array}{ll} a_1(x_1) - a_2(x_2) + a_1(x_1) \cdot a_2(x_2) & \text{if } x_1 < x_2 \\ a_1(x_1) - a_2(x_2) & \text{if } x_1 = x_2 \\ a_1(x_1) - a_2(x_2) - a_1(x_1) \cdot a_2(x_2) & \text{if } x_2 < x_1 \end{array}$$

Set α_i to be Player i's secure utility level and prove

$\alpha_1 < v < \alpha_2$ where v is the value of the noisy duel (Example 1).

Solution:

First compute $\alpha_1 = \sup\limits_{x_1} \inf \{ 2a_1(x_1) - 1, a_1(x_1) - a_2(x_1) - a_1(x_1) \cdot a_2(x_1) \}$. Next define $\check{t}$ by $a_1(\check{t}) + a_1(\check{t}) \cdot a_2(\check{t}) = 1$ and check that α_1 is worth $\alpha_1 = \sup\limits_{x \geq \check{t}} \{ a_1(x) - a_2(x) - a_1(x) a_2(x) \}$. Similarly $\alpha_2 = \inf\limits_{x \geq \check{t}} \{ a_1(x) - a_2(x) + a_1(x) \cdot a_2(x) \}$. Denote by x* the optimal strategy (of both players) in the noisy duel (Example 1). By definition of $\check{t}$ we have $a_1(\check{t}) + a_2(\check{t}) < 1$; hence $\check{t} < x^*$. We must prove $\alpha_1 < v$ where

$$\alpha_1 = \sup\limits_{x \geq \check{t}} m(x),$$

where $m(x) = a_1(x) - a_2(x) - a_1(x) \cdot a_2(x)$.

We prove first $x \geq x^* \Rightarrow m(x) < v$. Since $a_2(x^*) \leq a_2(x)$, we get

$$m(x) \leq a_1(x) - a_2(x^*) - a_1(x) a_2(x^*) = a_1(x^*) a_1(x) - a_2(x^*)$$

$$< a_1(x^*) - a_2(x^*) = v$$

We prove next $\check{t} \leq x \leq x^* \Rightarrow m(x) < v$. From $\check{t} \leq x$ follows $(a_1 + a_2 + a_1 \cdot a_2)(x) \geq 1$; hence, $m(x) \leq 2a_1(x) - 1$. So,

$$m(x) \leq 2a_1(x) - 1 \leq 2a_1(x^*) - 1 = v$$

Observe that for $x < x^*$ the right-hand inequality is strict, and for $\check{t} < x$ the left-hand one is. So, we have proved $m(x) < v$ for all $x \geq \check{t}$, and $\alpha_1 < v$ follows by continuity of m. The proof of $v < \alpha_2$ goes by a symmetrical argument.

4) Borel's model of poker continued

Generalize the computation of Example 2 where the second bid of Player 1 is worth \$a, $a \geq 1$, and that of Player 2 is worth \$b, $b \geq 1$. Discuss according to the sign of $a^2 + 2a - 2b$.

Solution:

As in Example 2

$$u(x_1, x_2) = -P(m_1 < x_1) + P(x_1 \leq m_1) \{P(m_2 < x_2) +$$

$$P(x_2 \leq m_2)[(1 + b)\pi(x_1, x_2) - (1 + a)(1 - \pi(x_1, x_2))]\}$$

$$= -x_1 + (1 - x_1)\{x_2 + (1 - x_2)[(2 + a + b)\pi(x_1, x_2) - (1+a)]\}$$

Setting $y_i = 1 - x_i$, $i = 1, 2$, this reduces to

$$-1 + 2y_1 + by_1y_2 - (1 + \frac{a + b}{2})y_1^2 \qquad \text{if } y_1 \leq y_2$$

$$u =$$

$$-1 + 2y_1 - (2 + a)y_1y_2 + (1 + \frac{a + b}{2})y_2^2 \quad \text{if } y_2 \leq y_1$$

Thus u is continuous, concave w.r.t. y_1 and convex w.r.t. y_2. Search for its saddle point(s): Check first that if $y_1 = 0$, then $(\partial u/\partial y_1)(0, y_2) = 2 + by_2 > 0$, so $(0, y_2)$ cannot be a saddle point since Player 1 gains by increasing y_1.

Similarly, if $y_2 = 0$, $y_1 > 0$, then $(\partial u/\partial y_2) = -(2 + a)y_1 < 0$; so, Player 2 gains by increasing y_2. Next, at (y_1, y_2), such that $0 < y_1 \leq y_2$, we have $(\partial u/\partial y_2) = by_1 > 0$; so, Player 2 gains by lowering y_2. Thus our saddle point(s) must be such that $0 < y_2 \leq y_1$. On the interval $[0, y_1]$, u is concave w.r.t. y_2, and it is continuously differentiable at y_1. So we must have $(\partial u / \partial y_2) = -(2 + a)y_1 + (2 + a + b)y_2 = 0$. On the interval $[y_2, 1]$, u is linear in y_1 so we must have

$$\frac{\partial u}{\partial y_1} = 2 - (2 + a)y_2 = 0 \quad \underline{or} \quad (y_1 = 1 \text{ and } 2 - (2 + a)y_2 \geq 0)$$

We have two possible solutions: $y_2 = (2/2 + a) \Rightarrow$
$y_1 = [2(2 + a + b)/(a + b)^2]$ $\underline{or}$ $y_1 = 1 \Rightarrow y_2 = (2 + a/2 + a + b)$.
In the former case we must check $y_1 \leq 1$; in the latter we must check $2 - (2 + a)y_2 \geq 0$. Thus the final solution.

If $a^2 + 2a - 2b \geq 0$, then

$$y^*_1 = \frac{2(2 + a + b)}{(2 + a)^2} \in [0, 1], \quad y^*_2 = \frac{2}{2 + a}, \quad v = - \frac{(a^2 + 2a - 2b)}{(2 + a)^2} \leq 0$$

If $a^2 + 2a - 2b \leq 0$ then

$$y^*_1 = 1, \quad y_2 = \frac{2 + a}{2 + a + b}, \quad v = - \frac{(a^2 + 2a - 2b)}{2(2 + a + b)} \geq 0$$

5) <u>Final offer arbitration</u> (Brams and Merrill [1983])

Player 2 must pay \$x to Player 1 where x is determined as follows. Each player proposes an amount x_i, i = 1, 2. Next, the referee picks whichever of x_1, x_2 is closest to his own opinion y. Both players are aware of the probability distribution from which y is drawn. This distribution has a positive continuous density f on $]-\infty, +\infty[$. Denoting by F the cumulative distribution of f, the following game is in order.

$$X_1 = X_2 =]-\infty, +\infty[$$

$$u(x_1, x_2) = x_1 \cdot F\left(\frac{x_1 + x_2}{2}\right) + x_2 \cdot (1 - F\frac{x_1 + x_2}{2}) \text{ if } x_1 \leq x_2$$

$$= x_2 \cdot F\frac{x_1 + x_2}{2} + x_1 \cdot (1 - F\frac{x_1 + x_2}{2}) \text{ if } x_2 \leq x_1$$

Solve the system of first-order conditions for a saddle point of u and show that is has two solutions (α, β) and (β, α). Prove that at a saddle point of u we must have $x_2 \leq x_1$, which leaves us with only one possible saddle point. Write the second-order conditions: $\partial^2 u/\partial x_1^2 \leq 0$ and $\partial^2 u/\partial x_2^2 \geq 0$, assuming f is derivable. Conclude by giving an example of f where our game has a value and one where it does not.

Solution:

At any outcome (x_1, x_2) such that $x_1 < x_2$, Player 1 improves upon his payoff by switching to $x'_1 = x_2$:

$$u(x_1, x_2) = x_2 + (x_1 - x_2) F(\frac{x_1 + x_2}{2}) < x_2 = u(x_2, x_2)$$

Thus a saddle point must be such that $x_2 \leq x_1$.

First-order conditions at an outcome $x_2 \leq x_1$:

8

$$\frac{\partial u}{\partial x_1} = (x_2 - x_1/2)\, f(x_1 + x_2/2) - F(x_1 + x_2/2) + 1 = 0$$

$$\frac{\partial u}{\partial x_2} = (x_2 - x_1/2)\, f(x_1 + x_2/2) + F(x_1 + x_2/2) = 0$$

Hence $F(x_1 + x_2/2) = (1/2)$, which determines $x_1 + x_2/2 = x^*$ since F is strictly increasing. Summing up the two equations we get $x_1 - x_2 = 1/f(x^*)$, whence the solution

$$x^*_1 = x^* + \frac{1}{2f(x^*)}$$

$$x^*_2 = x^* - \frac{1}{2f(x^*)}$$

Write now the second-order conditions:

$$\frac{\partial^2 u}{\partial x_1^2} = \frac{(x^*_2 - x^*_1)}{4}\, f'(x^*) - f(x^*) \le 0$$

$$\frac{\partial^2 u}{\partial x_2^2} = \frac{(x^*_2 - x^*_1)}{4}\, f'(x^*) + f(x^*) \ge 0$$

This reduces to $|f'(x^*)| \le 4f^2(x^*)$. An example where our pair (x^*_1, x^*_2) actually is a saddle point is the normal distribution with mean 3. There $x^* = 3$ and $f'(x^*) = 0$. Also $x^*_1 = 3 + (\sqrt{2}\pi/2)$, $x^*_2 = 3 - (\sqrt{2}\pi/2)$.

An example where (x^*_1, x^*_2) is not a saddle point is any distribution such that $|f'(x^*)| > 4f^2(x^*)$, e.g.,

$$f(x) = \begin{cases} 0, & x \leq -5 \\ \frac{1}{12}, & -5 \leq x \leq 0 \\ \frac{1}{12} + x^2, & 0 \leq x \leq 1/2 \end{cases}$$

(We leave f unspecified for $x \geq (1/2)$.) Here $x^* = (1/2)$, $f'(x^*) = 1$, $f(x^*) = (1/3)$.

b) *Games in extensive form: examples*

6) <u>Splitting the pile: Singleton loses</u>

Solve the variant of Example 4 where the player who must pick a singleton loses.

Solution:

The partition $N_1 \cup N_2$ of N satisfies the same properties as in Example 4 (namely the system on top of page 25) but the initial condition is different: here $1 \in N_2$ while in Example 4, $1 \in N_1$.

Therefore N_1 is the set of even positive integers; N_2 is the set of odd positive integers. Indeed any even pile can be split in two odd piles, whereas any splitting of an odd pile must leave one even subpile.

7) <u>Removing sticks</u>

Players successively remove one stick <u>or</u> two <u>adjacent</u> sticks. The initial position is

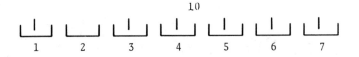

For instance, if Player 1 removes first 5, 6, Player 2 has 5
legal moves: remove 1 or 3 or 4 or 3, 4 or 7. Who wins
if the player removing the last stick loses? Who wins if
the player removing the last stick wins?

Solution:

Last stick loses, Player 2 wins.

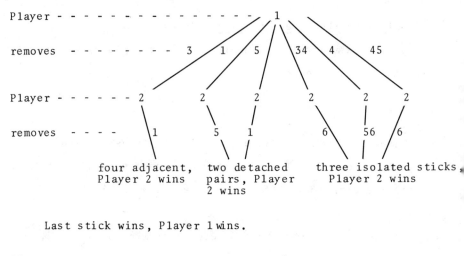

Last stick wins, Player 1 wins.

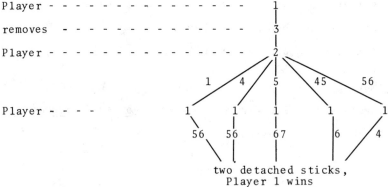

8) The Marienbad game

Physically the game amounts to a set of p rows of respectively n_1, ..., n_p sticks. Each player is asked to remove successively at least one and at most all sticks in exactly one row still alive. The player who picks the last stick loses.

We fix an integer p, $p \geq 1$ and for all p-tuples $\underline{n} = (n_1, ..., n_p)$ where n_1, ..., n_p are integers, possibly zero, we define two games, $G_{\underline{n}}^1$, $G_{\underline{n}}^2$, as follows. Given $\underline{n} \neq (0, 0, ..., 0)$, call $\underline{n}'$ a successor of $\underline{n}$ if there exists k, $1 \leq k \leq p$ such that

$$n'_{k'} = n_{k'}, \quad \text{for all } k', \ 1 \leq k' \leq p \quad \text{and} \quad k' \neq k$$

$$0 \leq n'_k < n_k$$

If $\underline{n} = (0, ..., 0)$, then in $G_{\underline{n}}^i$ Player i wins. If $\underline{n} \neq (0, ..., 0)$, then $G_{\underline{n}}^i$ is played as follows. First, Player i picks a successor $\underline{n}'$ of $\underline{n}$. If $\underline{n}' = (0, ..., 0)$, then Player j, $j \neq i$, wins the play.

Otherwise, the game $G_{\underline{n}}^j$ starts. Player j picks a successor $\underline{n}''$ of $\underline{n}'$. If $\underline{n}'' = (0, ..., 0)$ Player i wins.

Otherwise $G_{\underline{n}''}^i$ starts. And so on.

a) Prove the existence of a partition, $N_1 \cup N_2$ of $\{0, 1, 2, ...\}^p$ such that if $\underline{n}$ belongs to N_i, Plater i can force a win in $G_{\underline{n}}^1$.

b) For all $\underline{n} = (n_1, \ldots, n_p)$ denote

$$n_k = \alpha^k_\ell \alpha^k_{\ell-1}, \ldots, \alpha^k_1 \alpha^k_0, \qquad 1 \leq k \leq p$$

the binary representation of n_k where ℓ is an upper bound of the number of digits needed for $n_1, \ldots, n_p$. (Thus some, not all, among $\alpha^1_\ell, \ldots, \alpha^p_\ell$ might be zero.) Next denote

$$a_o = \sum_{k=1}^{p} \alpha^k_o, \ldots, a_\ell = \sum_{k=1}^{p} \alpha^k_\ell$$

Then prove that $N_2 = M_2 \cup P_2$ where

$M_2 = \{\underline{n}/ \ \forall \ j, \ 0 \leq j \leq \ell, \ a_j \text{ is even and } \exists j, \ 1 \leq j \leq \ell, \ a_j > 0\}$

$P_2 = \{\underline{n}/ \ \forall \ j, \ 1 \leq j \leq \ell, \ a_j = 0 \text{ and } a_o \text{ is odd}\}$

Solution:

The existence of the partition follows again from Kuhn's theorem. We prove that the proposed partition is indeed the correct one. Check first the initial conditions: If $\underline{n}$ contains exactly one row alive, i.e., exactly one non-zero n_k, then Player 1 wins if $n_k > 1$ and loses if $n_k = 1$. Indeed, if $n_k > 1$, $\underline{n}$ belongs neither to N_2 nor to P_2, and if $n_k = 1$, $\underline{n}$ is in P_2.

We check now that from any $\underline{n} \ \varepsilon \ N_2$, Player 1 $\underline{must}$ go to some position in N_1 while from any $\underline{n} \ \varepsilon \ N_1$ Player 1 $\underline{can}$ go to a position in N_2.

First, if $\underline{n} \ \varepsilon \ N_2$, any legal move puts us in N_1. Suppose $\underline{n} \ \varepsilon \ M_2$ and we reduce pile k. Let α^k_j be the leftmost 1 in n_k's binary representation. Then a_k must reduce by exactly 1. Next if $\underline{n} \ \varepsilon \ P_2$, it is trivial that any move takes us to N_1.

Now if $\underline{n} \ \varepsilon \ N_1$, there is a possible move to N_2. This is obvious if $a_j = 0$, all $j = 1, \ldots, \ell$ and a_0 is even. Otherwise let $j^* = \max \{j : a_j$ is odd$\}$. For some k, $\alpha^k_{j^*} = 1$. Reduce n_k by changing $\alpha^k_{j^*}$ to zero and choosing α^k_j, for all $j < j^*$ so that a_j is even if $j \geq 1$ and a_0 is odd. Of course, we may have to increase (from zero to 1) some α^k_j, $j < j^*$, but n_k decreases since $\alpha^k_{j^*}$ does and α^k_j, $j > j^*$ remain fixed.

c) *Abstract normal form games*

9) <u>Symmetrical games</u>

A two-person, zero-sum game is symmetrical if $X_1 = X_2 = X$ and moreover

$$u(x_1, \ x_2) = -u(x_2, \ x_1) \qquad \text{all } x_1, \ x_2 \ \varepsilon \ X$$

Prove that the value of a symmetrical game (if any) is zero and optimal strategies of both player coincide. What if the game has no value?

Solution:

$$\alpha_2 = \inf_{x_2} \sup_{x_1} u(x_1, \ x_2) = \inf_{x_1} \sup_{x_2} u(x_2, \ x_1) = \inf_{x_1} \sup_{x_2} - u(x_1, \ x_2)$$

$$= \inf_{x_1} (-\inf_{x_2} u(x_1, \ x_2)) = -\sup_{x_1} \inf_{x_2} u(x_1, \ x_2) = -\alpha_1$$

14

If x_1^* is optimal for Player 1: $\inf\limits_{x_2} u(x_1^*, x_2) = \alpha_1$, then

$-\alpha_1 = -\inf\limits_{x_2} - u(x_2, x_1^*) = \sup\limits_{x_1} u(x_1, x_1^*)$, so it
is optimal for Player 2 as well.

10) Shapley's criterion for finite games

Let $G = (X_1, X_2, u)$ be a finite game (both X_1, X_2 are finite) such that for all doubletons $Y_1 \subset X_1$, $Y_2 \subset X_2$ the restricted game (Y_1, Y_2, u) has a value. Show that G has a value.

Solution:

Assume G has no value. Without loss, assume $\alpha_1 < 0 < \alpha_2$. Pick a strategy $x_2^* \in X_2$ which maximizes the integer-valued function :

$$\rho(x_2) = \# \{x_1 \in X_1 | u(x_1, x_2) \le 0\}$$

Given x_2^*, we have $\sup\limits_{x_1} u(x_1, x_2^*) \ge \alpha_2 > 0$; hence we can pick x_1^* such tat $u(x_1^*, x_2^*) > 0$. Similarly, $\inf u(x_1^*, x_2) \le \alpha_1 < 0$, so we choose $\tilde{x}_2$ such that $u(x_1^*, \tilde{x}_2) < 0$. By construction of x_2^* we have $\rho(\tilde{x}_2) \le \rho(x_2^*)$. On the other hand,

$$u(x_1^*, \tilde{x}_2) < 0 < u(x_1^*, x_2^*)$$

By definition of ρ this implies the existence of $\tilde{x}_1$ such that

$$u(\tilde{x}_1, x_2^*) \le 0 < u(\tilde{x}_1, \tilde{x}_2)$$

Now consider the restriction of our game to $Y_1 \times Y_2 = \{x^*_1, \tilde{x}_1\} \times \{x^*_2, \tilde{x}_2\}$. This game has no value.

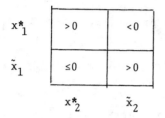

	x^*_2	$\tilde{x}_2$
x^*_1	> 0	< 0
$\tilde{x}_1$	≤ 0	> 0

11) Proof of Remark 1

Let $G = (X_1, X_2, u)$ be a two-person, zero-sum game where X_i, $i = 1, 2$ are arbitrary sets. Define x^*_1 to be an ε-prudent strategy of Player 1 iff $u(x^*_1, x_2) \geq \alpha_1 - \varepsilon$ for all x_2 and x^*_2 to be an ε-prudent strategy of Player 2 iff $u(x_1, x^*_2) \leq \alpha_2 + \varepsilon$ for all x_1. Similarly, define and ε saddle point by

$$u(x_1, x^*_2) - \varepsilon \leq u(x^*_1, x^*_2) \leq u(x^*_1, x_2) + \varepsilon, \text{ for all } x_1,$$

x_2. Throughout the exercise we assume that u is uniformly bounded on $X_1 \times X_2$. There exist two real numbers a, b such that

$$a \leq u(x_1, x_2) \leq b \qquad \text{all } (x_1, x_2) \in X_1 \times X_2$$

Prove the following claims.

a) For any $\varepsilon > 0$, each player has an ε-prudent strategy in G. b) If G has a value and $x^* = (x^*_1, x^*_2)$ is a pair of ε-prudent strategies, then x^* is a 2ε saddle point. c) If x^* is an ε saddle point, then x^*_i is a 2ε prudent strategy of Player i (for i = 1, 2) and $(\alpha_2 - \alpha_1) \leq 2\varepsilon$. d) If, for all

$\epsilon > 0$, G has an ϵ saddle point, then G has a value.

Solution:

(a) By definition of a supremum, $\alpha_1 = \sup\limits_{x_1 \epsilon X_1} [\psi(x_1)]$ where

$\psi(x_1) = \inf\limits_{x_2 \epsilon X_2} u(x_1, x_2)$ implies

$$\forall \epsilon > 0 \; \exists \; x_1^* \; \epsilon \; X_1 \quad \psi(x_1^*) \geq \alpha_1 - \epsilon$$

Such an x_1^* is an ϵ-prudent strategy for Player 1. The argument for Player 2 is symmetrical.

(b) Assume $\alpha_1 = \alpha_2 = v$ and x_i^* is ϵ prudent for Player i. Then,

$$u(x_1^*, x_2) \geq v - \epsilon \qquad \text{all } x_2 \qquad (1)$$

in particular

$$u(x^*) \geq v - \epsilon \qquad (2)$$

Also,

$$u(x_1, x_2^*) \leq v + \epsilon \qquad \text{all } x_1 \qquad (3)$$

in particular

$$u(x^*) \leq v + \epsilon \qquad (4)$$

$(1) + (4) \Rightarrow u(x^*) \leq u(x_1^*, x_2) + 2\epsilon \qquad \text{all } x_2$

$(2) + (3) \Rightarrow u(x_1, x_2^*) - 2\epsilon \leq u(x^*) \qquad \text{all } x_1$

(c) Say x^* is an ε saddle point. This does not imply that G has a value. By definition of an ε saddle point:

$$\sup_{x_1} u(x_1, x_2^*) - \varepsilon \leq u(x^*) \leq \inf_{x_2} u(x_1^*, x_2) + \varepsilon$$

Hence $\alpha_2 - \varepsilon \leq u(x^*) \leq \alpha_1 + \varepsilon$. (Since $\alpha_1 \leq \alpha_2$, this implies $(\alpha_2 - \alpha_1) \leq 2\varepsilon$). Next we have

all x_1, $u(x_1, x_2^*) - \varepsilon \leq u(x^*) \Rightarrow u(x_1, x_2^*) \leq \alpha_1 + 2\varepsilon$ all x_1

all x_2, $u(x^*) \leq u(x_1^*, x_2) + \varepsilon \Rightarrow u(x_1^*, x_2) \geq \alpha_2 - 2\varepsilon$ all x_2

That is, x_i^* is 2ε prudent, $i = 1, 2$.

(d) We know from question (c) that if G has an ε saddle point, then $\alpha_2 - \varepsilon \leq \alpha_1 + \varepsilon \leq \alpha_2 + \varepsilon \Rightarrow \alpha_2 - \alpha_1 \leq 2\varepsilon$. This holds for all $\varepsilon > 0$; hence, $\alpha_2 = \alpha_1$.

12) A topological duel (Choquet)

Let E be a metric space. We denote by zero the set of subsets of E with a nonempty interior. Our game works as follows: In Step 1, Player 1 picks an $A_1 \, \varepsilon \, 0$. In Step 2, Player 2 picks an $A_2 \, \varepsilon \, 0$ with the only constraint $A_2 \subset A_1$....
In Step t a player (1 for odd t, 2 for even t) picks an $A_t \, \varepsilon \, 0$ with the only constraint $A_t \subset A_{t-1}$ and so on undefinitely. We say that Player 1 wins the play if

$$\bigcap_{t=1}^{\infty} A_t \neq \emptyset$$

If this intersection is empty, we say that Player 2 wins the play. a) Prove that if E is a complete metric space, Player 1 can force a win. b) Prove that if E = Q (the rational numbers), Player 2 can force a win. c) Are the results affected if both players must now pick nonempty open subsets of E?

Solution:

(a) Player 1 picks closed balls whose radius decreases at least by 1/2 after each pair of moves (one for each player). Since these balls are inclusion decreasing and E is complete, their intersection is nonempty.

(b) For all integers p, let $Q(p) = \{(n/p)/n \; \varepsilon \; Z\}$ where Z is the set of positive or negative integers. Player 2 wins by picking as his pth move an interval which does not intersect $Q(p)$. Since Q is the union of all $Q(p)$, p = 1, 2, ..., the overall intersection has to be empty.

(c) Player 1 picks now open balls B_1, B_2, ..., the radius of which decreases by at least 1/2 and such that, in addition, the closure $\overline{B}_{t+1}$ of B_{t+1} is contained in B_t. Henceforth $\underset{t}{\cap} B_t = \underset{t}{\cap} \overline{B}_t$, and we are back to question (a). In question (b) it is plain that Player 2 can choose open intervals with the required property.

CHAPTER 2   TACTICAL EXCHANGES OF INFORMATION

1)   A counterintuitive noninessential game

Give an example of a two-person game which is <u>not</u>
inessential but with a Pareto-optimal outcome x such that
$\beta_i = u_i(x)$, i = 1, 2.

<u>Note</u>: Outcome x is Pareto-optimal if there is no outcome
y such that $u_i(x) \leq u_i(y)$, i = 1, 2 and at least one
inequality is strict.

Solution:

1	2
2	1
0	0
0	3

Here $\beta_1 = 1$, $\beta_2 = 2$, yet $\alpha_1 = 1$, $\alpha_2 = 1$.

2) A consequence of the competition for the second move

If the competition for the second move arises in game (X_1, X_2, u_1, u_2), prove that neither game $(X_1, X_2, u_1, -u_1)$ nor $(X_1, X_2, -u_2, u_2)$ has a value.

Solution:

Suppose $(X_1, X_2, u_1, -u_1)$ has a value. Take a prudent strategy x_1 for Player 1 and a best reply x_2 of Player 2 to x_1. Then,

$$u_1(x_1, x_2) \geq \inf_{y_2} u_1(x_1, y_2) = \alpha_1 = \beta_1$$

$$u_2(x_1, x_2) = \sup_{y_2} u_2(x_1, y_2) \geq \beta_2$$

Hence the competition for the second move does not arise. Note that we had to assume existence of prudent strategies and best replies: This holds if X_1, X_2 are compact and u_1, u_2 are continuous.

3) On games where each player has a value

Give an example of a 2 x 2 game where an outcome $x^* = (x^*_1, x^*_2)$ is a saddle point of u_1 and of $-u_2$:

$$u_1(x_1, x^*_2) \leq u_1(x^*_1, x^*_2) \leq u_1(x^*_1, x_2)$$

$$u_2(x^*_1, x_2) \leq u_2(x^*_1, x^*_2) \leq u_2(x_1, x^*_2)$$

yet the game is not inessential.

Solution:

Prisoner's dilemma .

4) Picking numbers

Both players pick an integer x_i, $1 \le x_i \le 10$. If $x_1 + x_2 = 10$ then Player i's payoff is x_i. Otherwise, the payoff is $(4, 0)$ if $x_1 + x_2$ is even and $(0, 4)$ if $x_1 + x_2$ is odd.

Show that the competition for the first move arises in this game. However, replacing the altruistic assumption by the opposite assumption in the definition of the utility levels S_i (so that a player, when indifferent towards choosing two best-reply strategies, chooses so as to minimize his opponent's utility) is enough to avoid this competition.

Solution:

Compute Player 1's Stackelberg utility level S_1. To any x_1, $0 \le x_1 \le 5$ the best reply of Player 2 is $x_2 = 10 - x_1$. Thus $S_1 \ge 5$. To $x_1 = 6$, Player 2 has several best replies: $x_2 = 4$ (yielding $u_1 = 6$) or x_2 odd (yielding $u_1 = 0$). To $x_1 \ge 7$ Player 2's best reply is to force payoffs $(0, 4)$. Therefore $S_1 = 6$ under the optimistic definition, whereas $S'_1 = 5$ under the pessimistic one. By symmetry $S_2 = 6$, $S'_2 = 5$ as well. The vector $(6, 6)$ is not a feasible payoff vector, but $(5, 5)$ is.

22

5) <u>Strategic voting</u>: <u>Borda versus Rawls</u>

There are seven candidates a, b, c, d, e, f, g and two voters. A ballot of Voter i, i = 1, 2 is a linear ordering such as

$$x_i = bdgecaf$$

      ↑   ↑

    top   bottom

   || ||

best preferred least preferred

Given two ballots (x_1, x_2) the elected candidate would turn out to be as follows.

<u>Borda</u>

Give six points for the top candidate in any ballot, five points to the second best and so on. Then add both scores to find a candidate's Borda score. Break ties by lexicographic ordering. For instance,

x_1	cgdebaf		abcdefg
x_2	bafgdce	Borda score	6876348

Hence b is elected. We denote by $B(x_1, x_2)$ this voting rule.

<u>Rawls</u>

Define the score of a candidate as the smallest of the two scores. Again break ties lexicographically. In the example above

	abcdefg
Rawls Score	1212003

g is elected. Denote $R(x_1, x_2)$ for this voting rule.

Voters cast their ballots strategically, hence they are not necessarily truthful. Let u_i be Voter i's true preference, written as a fixed-scale utility function with range $\{0, \ldots, 6\}$ (hence voters are never indifferent about any two candidates).

The two voting rules described above generate two games

$$(X_1, X_2, u_1 B, u_2 B)$$

$$(X_1, X_2, u_1 R, u_2 R)$$

where $X_1 = X_2$ is the set (with cardinality 7!) of all linear orderings of our seven candidates.

a) Show that

$$\sup_{x_i} \inf_{x_j} u_i B(x) \le 1 \le 5 \le \inf_{x_j} \sup_{x_i} u_i B(x).$$

Deduce that, in general, competition for the second move arises in the Borda game. Make clear when it does not arise.

b) Show that

$$\sup_{x_i} \inf_{x_j} u_i R(x) = \inf_{x_j} \sup_{x_i} u_i R(x) = 3$$

and deduce that the competition for the second move never

arise in the Rawls game. Next show that Player i's optimal utility as a leader is

$$S_i = \sup \{u_i(z) \,|\, u_j(z) \geq 3\}$$

and deduce that in general the competition for the first move arises in the Rawls game (again, give the exact meaning of "in general").

Solution:

(a) If Player i casts any ballot x_i, e.g.,

$$x_i \; : \; \text{bcagefd}$$

he can at most guarantee that d is not elected because Player j has a strategy x_j that forces election of anyone among b, c, a, g, e, f given x_i. For instance to elect f, take x_j as follows:

$$x_j \; : \; \text{fdegacb}$$

On the other hand, d will never be elected no matter what x_j. Indeed the only ballot by Player j where d has maximal score is the reverse ballot dfegacb whereupon all candidates tie and a is elected (ties are broken lexiographically).

Thus by ranking candidate $z \neq a$ at the bottom of his ballot, a player effectively eliminates z. However, a can never be eliminated. Given any ballot x_i' where a is bottom, the reverse ballot $\bar{x}'_i$ forces election of a. Thus equality sup inf $u_i B = 1$ holds only for those u_i of which a is not bottom. When a is bottom of u_i, we have sup inf $u_i B = 0$.

The inequality $5 \leq \inf_{x_j} \sup_{x_i} u_i B$ follows from the previous argument when applied to the reverse utility $\bar{u}_i = 6 - u_i$ while exchanging players (which is possible since the voting rule is symmetrical):

$$1 \geq \sup_{x_j} \inf_{x_i} \bar{u}_i = 6 - \inf_{x_j} \sup_{x_i} u_i$$

The competition for the second move must arise if (5, 5) is not a feasible utility vector, that is to say if the two sets of the two best preferred candidates of each player do not intersect. Given the above remark that inf sup $u_i B$ is 5, unless a is top of u_i, in which case it is worth 6, we conclude:

· If neither player has a for top, competition for
  the second move arises unless there is a candidate
  ranked top or second for both.
· If one player's top is a, competition for the second
  move arises unless the other ranks a top or second.

(b)   First by casting his true preferences $x_i = u_i$ a player guarantees utility 3:

$$u_i R(u_i, x_j) \geq 3 \qquad \qquad \text{all } x_j$$

Namely, if u_i = bdgecaf, say, then at least one among bdge must be ranked among the four first candidates of x_j; hence, they will get a Rawls score of at least 3. But c, a, f get a strictly lower Rawls score.

Thus $\sup_{x_i} \inf_{x_j} u_i R \geq 3$, all u_i. Next consider $\bar{u}_i = 6 - u_i$, the reverse ballot of some arbitrary u_i. By symmetry of R in its two variables,

$$\sup_{x_j} \inf_{x_i} \bar{u}_i R = \sup_{x_i} \inf_{x_j} \bar{u}_i R \geq 3$$

However,

$$\sup_{x_j} \inf_{x_i} \bar{u}_i R = 6 - \inf_{x_j} \sup_{x_i} u_i R$$

Thus $\inf_{x_j} \sup_{x_i} u_i R \leq 3$. Together with $\sup_{x_i} \inf_{x_j} u_i R \geq 3$, this proves the desired equality.

We compute next a player's Stackelberg utility. Suppose Player 1 is leader and that Player 2's utility is u_2:

$$u_2 : \text{dgbaecf}$$

We know that Player 2 is guaranteed that the elected candidate z is in {d, g, b, a}, i.e., $u_2(z) \geq 3$. The point is that Player 1 as a leader can force the election of _any_ candidate in {d, g, b, a} by putting the remaining three at the bottom of his ballot x_1^*. For instance, to elect b, he reports x_1^* : xxxxdga. Given x_1^* any best reply x_2^* of Player 2 guarantees utility level 3 to Player 2:

$$u_2 R(x_1^*, x_2^*) = \sup_{x_2} u_2 R(x_1^*, x_2) \geq 3 \quad \Rightarrow \quad R(x_1^*, x_2^*) = b$$

This proves the desired formula for S_i.

Hence, competition for the first move arises <u>unless</u> there is a candidate z such that

$$u_1(z) < u_1(z') \quad \Rightarrow \quad u_2(z') < 3 \qquad \text{all } z'$$

$$u_2(z) < u_2(z') \quad \Rightarrow \quad u_1(z') < 3 \qquad \text{all } z'$$

A few trials will convince the reader that this is not a frequent situation.

CHAPTER 3   DOMINATING STRATEGIES

a)   *Examples of games with dominating strategy equilibrium*

1)   <u>A quantity setting oligopoly</u> (Case [1979])

Suppose that the price of a typical satiable good, say mineral water, goes to ce^{-S} where S is the total supply. If n costless producers control the quantities x_1, ..., x_n of mineral water that they supply, we obtain the following game

$$u_i(x_1, \ldots, x_n) = cx_i e^{-(x_1 + \ldots + x_n)}$$

Compute the equilibrium in dominating strategies and comment upon the assumed form of the inverse demand.

Solution:

Since xe^{-x} reaches its maximum at $x = 1$, the strategy $x_i = 1$ is dominating for all players. The demand is a multiplicative function of the individual supplies, so the elasticity of the demand w.r.t. firm i's supply is independent of that of firm j's.

2) <u>A price setting duopoly</u> (Case [1979])

Two duopolists offer two substitutable goods. If they set the prices p_1, p_2 the corresponding demands are

$$d_1 = \left(\frac{p_2}{p_1}\right)^{\alpha_1} \quad \text{units of the good produced by Player 1}$$

$$d_2 = \left(\frac{p_1}{p_2}\right)^{\alpha_2} \quad \text{units of the good produced by Player 2}$$

We assume $\alpha_i \geq 1$, $i = 1, 2$. Suppose, in addition, a constant return to scale technology for both producers. Thus the following normal form game.

$$u_i(p_1, p_2) = (p_i - c_i) \cdot d_i$$

where c_i is constant.

Compute the equilibrium in dominating strategies of this game and comment on the assumed form of the demand functions.

Solution:

As

$$u_1(p_1, p_2) = p_2^{\alpha_1} \left(\frac{p_1 - c_1}{p_1^{\alpha_1}}\right)$$

the maximum of u_1 at every positive level p_2 is reached at
$p_1^* = (\alpha_1/\alpha_1 - 1) \cdot c_1$, so,

$$\left(\frac{\alpha_1}{(\alpha_1 - 1)} c_1, \frac{\alpha_2}{(\alpha_2 - 1)} c_2\right)$$

is the equilibrium in dominating strategies. As in Exercise 1,
the key is the multiplicative character of the demand function:
$(\partial d_i/d_i)/(\partial p_i/p_i)$ is independent of p_j.

3) A Colonel Blotto game

There are ten locations with respective value $a_1 < \ldots < a_{10}$.
Player i (i = 1, 2) is endowed with n_i soldiers ($n_i < 10$)
and must allocate them among the locations. To each particular
location he can allocate no more than one soldier. The
payoff at location p is a_p to the player whose soldier is
unchallenged, and $-a_p$ to his opponent, unless both have a
soldier at p or no one has, in which case the payoff is
zero to both. The total payoff is obtained by summing up
local payoffs.

Show that this game has a unique dominating strategy
equilibrium. What if some of the a_p coincide?

Solution:

Pick any two locations p, p' such that p < p'; therefore $a_p < a_{p'}$. Any strategy x_i where Player i has a soldier on k and none on k' is dominated by x'_i , obtained by switching that soldier from k to k' (everything else being unchanged). Thus Player i's dominating strategy is to occupy the n_i most valuable positions. If some a_p coincide, this creates some equivalent strategies. Still a strategy is dominating if and only if it occupies n_i most valuable positions.

4) Games with linear payoffs (Moulin [1979])

a) Consider first a two-player game with strategy sets $X_1 = X_2 = [-1, +1]$ (the real interval) and linear utility functions

$$u_1(x_1, x_2) = ax_1 + bx_2, \quad u_2(x_1, x_2) = cx_1 + dx_2$$

where a, b, c, d are four fixed real numbers. Clearly each player has a dominating strategy (unique if a — resp d — is nonzero).

We are interested in configurations where a prisoner's dilemma arises, namely, the dominating strategy equilibrium is Pareto dominated. Prove that this is the case iff

$$a.c < 0, \quad b.d < 0, \quad \text{and} \quad 1 < \frac{b.c}{a.d}$$

Hint: Draw a picture of the feasible utility set, namely
$(u_1, u_2)(X_1 \times X_2)$.

   b) Now we have a n-person game with $X_i = [-1, +1]$,
all i = 1, ..., n, and Player i's utility function

$$u_i(x) = \sum_{j=1}^{n} a^j_i x_j$$

We assume $a^i_i \neq 0$, all i = 1, ..., n. Denote by x* the
unique dominating strategy equilibrium. Prove the equivalence
of the two following statements:

   i) Prisoner's dilemma effect: x* is a Pareto-
       dominated outcome.

   ii) There is an outcome $y \in X_N$ such that

$$\{a^i_i y_i > 0 \quad \text{and} \quad u_i(y) < 0\} \quad \text{all i = 1, ..., n}$$

       This says that each player would be better off
       by using y_i alone (i.e., while others stay put
       at strategy zero), but all will be worse off
       when they all use y_i.

   c) Example. Consider the game "provision of a public
good":

$$u_i = -\alpha_i x_i + \lambda \{ \sum_{j=1}^{n} x_j \} \qquad \alpha_i > 0$$

Interpret x_i as the (positive or negative) effort spent
by i for the common welfare, and $-\alpha_i$ as his marginal disutility
for the effort. For which values of λ do we have a prisoner's
dilemma effect?

Solution:

(a) If $a \neq 0$, clearly $D_1(u_1) = \{a/|a|\}$, whereas, if $a = 0$, $D_1(u_1) = X_1$ (all Player 1's strategies are equivalent-- to him). Similarly, if $d \neq 0$, $D_2(u_2) = \{d/|d|\}$ and if $d = 0$, $D_2(u_2) = X_2$. We seek to characterize situations where $D_1(u_1) \times D_2(u_2)$ does not contain any Pareto optimal outcome. This is possible only if both a, d are nonzero.

Assume, without loss of generality, $a > 0$, $d > 0$, so that $D_1(u_1) \times D_2(u_2) = \{(1, 1)\}$. Write that $(1, 1)$ is Pareto dominated

$$\exists \ x_1, \ x_2 \ \epsilon \ [-1, +1] \quad a + b \le ax_1 + bx_2$$

$$c + d \le cx_1 + dx_2$$

with at least one strict inequality. This is equivalent to:

$$\exists \ y_1, \ y_2 \ge 0 \quad ay_1 + by_2 \le 0$$

$$cy_1 + dy_2 \le 0$$

with at least one strict inequality. Clearly y_1 and y_2 must be both positive (non-zero) hence this system is equivalent to

$$0 < \frac{y_1}{y_2} \le - \frac{b}{a} \ , \qquad 0 < \frac{y_2}{y_1} \le - \frac{c}{d}$$

$$\uparrow \text{\rule{0pt}{0pt}} \qquad \text{at least one strict} \qquad \uparrow$$

which is possible iff b, c < 0 and $[-(b/a)] \cdot [-(c/d)] > 1$; i.e., b, c < 0 and ad < bc.

If a and/or d is negative, the corresponding inequalities are

$$a \cdot d < 0, \quad c \cdot d < 0 \quad \text{and} \quad ad \cdot (ad - bc) < 0$$

One typical example where this prisoner's dilemma effect arises is

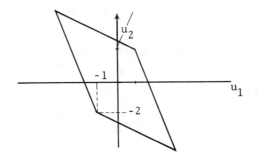

Below is the picture of the corresponding feasible utility set

$$
\begin{array}{|cc|}
\hline
a & b \\
\\
c & d \\
\hline
\end{array}
=
\begin{array}{|cc|}
\hline
1 & -2 \\
\\
-3 & 1 \\
\hline
\end{array}
$$

the dominant strategy equilibrium yields utilities (-1, -2).

(b) Denote A the matrix

$$[a^j_i]$$
$$1 \le i, j \le n$$

where i is the row index. Denote e the dominant strategy

equilibrium, namely the vector $e_i = (a_i^i / |a_i^i|)$ all $i = 1, \ldots, n$.

Next denote $C = [-1, +1]^n$. Finally the notation $x \ll y$ stands for $x_i < y_i$, all $i = 1, \ldots, n$.

In matrix notations our two conditions write as:

i) $\exists \; x \; \epsilon \; C \quad Ae \ll Ax$

ii) $\exists \; y \; \epsilon \; C \quad y_i \cdot e_i > 0$ all i, and $Ay \ll 0$

If (i) holds, set $y = (e - x/2)$; if (ii) holds, set $x = e - y$. Next observe that (ii) is equivalent to

$$\exists \; y \; \epsilon \; R^n \quad y_i \cdot e_i > 0 \qquad \text{all } i, \text{ and } Ay \ll 0$$

(c) By question b, a Prisoner's dilemma arises iff we can find $y_i < 0$, $i = 1, \ldots, n$ such that

$$-\alpha_i y_i + \lambda \left(\sum_{j=1}^{n} y_j \right) < 0$$

Setting $z_i = y_i / \sum_{j=1}^{n} y_j$, $z_i > 0$, this is equivalent to:

$z_i < \lambda/\alpha_i$, $i = 1, \ldots, n$. It is possible to find such a vector z iff

$$\sum_{i=1}^{n} \frac{\lambda}{\alpha_i} > 1 \iff \sum_{i=1}^{n} \frac{1}{\alpha_i} > \frac{1}{\lambda}$$

b) *Abstract games*

5) <u>On games with Pareto-optimal dominating strategy equilibrium</u>

Give an example of a two-person game with Pareto-optimal dominating strategy equilibrium, but which is <u>not</u> inessential.

Solution:

top

1 1	0 0
0 0	-1 -1

left

(top, left) is the dominating strategy equilibrium. Yet
$\alpha_1 = \alpha_2 = 0$.

6) <u>On inessential games and dominated strategies</u>

Give an example of a 3 x 3 two-person game (i.e., each
player has exactly three strategies) where no strategy is
dominated and no two are equivalent for any player ($\mathcal{D}_i = X_i$,
i = 1, 2) yet the game <u>is</u> inessential. Can you find a similar
example with a 2 x 3 or a 2 x 2 game?

Solution:

1 1	2 0	2 0
0 0	3 0	0 3
0 2	0 3	3 0

$\alpha_1 = \alpha_2 = 1$

With a 2 x 2 game such a configuration is not possible. Say $\alpha_1 = \alpha_2 = 0$, and the game is

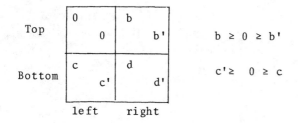

$$b \geq 0 \geq b'$$

$$c' \geq 0 \geq c$$

We must have $d \leq 0$ and/or $d' \leq 0$. Otherwise (α_1, α_2) is Pareto dominated. If $d \leq 0$, then top dominates bottom, unless $b = 0$ and $c = 0$, in which case top and bottom are equivalent. A similar argument shows that the desired configuration cannot arise in 2 x 3 games either.

7) <u>Games with equivalent undominated strategies</u>

Recall that strategies x_i and y_i of Player i are <u>equivalent</u> if they are not distinguishable in Player i's opinion:

$$\forall\, x_{-i} \in X_{-i}: \quad u_i(x_i, x_{-i}) = u_i(y_i, x_{-i})$$

Suppose that game $(X_j, u_j; j = 1, \ldots, n)$ is finite or X_j are compact and u_j continuous. Then the three following properties are equivalent. i) A dominating strategy for Player i exists: $D_i(u_i) \neq \emptyset$, ii) All strategies in $\mathcal{D}_i(u_i)$ are equivalent, iii) $D_i(u_i) = \mathcal{D}_i(u_i)$. Deduce that a player has a unique undominated strategy if and only if he has a unique dominating strategy.

Solution:

i) => (iii).

Take $x_i \in D_i$ and $y_i \in \mathcal{D}_i$. Then clearly x_i and y_i are equivalent; hence $y_i \in D_i$. Thus, $\mathcal{D}_i \subset D_i$, while the converse inclusion is obvious

iii) => ii)

Two dominating strategies are clearly equivalent

ii) => i)

Take any $x_i \in \mathcal{D}_i$; if $x_i \notin D_i$, there exists y_i, x_{-i} such that

$$u_i(x_i, \, x_{-i}) < u_i(y_i, \, x_{-i})$$

By the topological assumptions, there exist a strategy $z_i \in \mathcal{D}_i$ such that $u_i(y_i, \, x_{-i}) \le u_i(z_i, \, x_{-i})$. This implies that x_i, z_i are not equivalent. Contradiction.

8) <u>Topological properties of D_i and $\mathcal{D}_i$</u>

For all $i = 1, \ldots, n$, let X_i be a compact set and u_i be a continuous function on X. Show that the sets $\mathcal{D}_i(u_i)$ of undominated strategies are not necessarily closed. What about the sets $D_i(u_i)$ of undominating strategies?

<u>Hint</u>: Consider the following game.

$$X_1 = X_2 = [0, \, 1] \quad u_1(x_1, \, x_2) = 0 \quad \text{if } \frac{x_1}{2} + x_2 \ge 1$$

$$u_1(x_1, \, x_2) = x_1 - \frac{x_1 x_2}{1 - \frac{x_1}{2}} \qquad \text{if } \frac{x_1}{2} + x_2 \ge 1$$

u_2 is arbitrary.

Prove that u_1 is continuous and yet $\mathcal{D}_1(u_1)$ is not closed.

Solution:

The set $D_i(u_i)$ is closed as defined by a (possibly infinite) set of inequalities $\geq$.

In the example one checks that u_i is continuous. More-over, $\mathcal{D}_1(u_1) = \,]0, 1]$. Notice first

$$u_1(0, x_2) = 0$$

while

$$u_1(1, x_2) = \begin{cases} 0 & \text{if } \frac{1}{2} \leq x_2 \\ 1 - 2x_2 & \text{if } 0 \leq x_2 \leq \frac{1}{2} \end{cases}$$

Thus zero is dominated by 1. Next check that any $x_1 > 0$ is not dominated by $x'_1 > x_1$, since for $x_2 \in \,]1 - \dfrac{x'_1}{2}, 1 - \dfrac{x_1}{2}[$ we have $u_1(x_1, x_2) > 0 = u_1(x'_1, x_2)$. Also $x_1 > 0$ is not dominated by $x'_1 < x_1$ since, for all $x_2 < 1 - (x_1/2)$, we have $u_1(x_1, x_2) > u_1(x'_1, x_2)$.

CHAPTER 4   SOPHISTICATED AND PERFECT

EQUILIBRIUM

a) *Examples*

1) <u>A two-step competition for the first move</u>

Consider the following two-person game. <u>Round 1</u>: Player 1 calls or folds. If he folds, the game is over and the payoffs are (0, 1). If he calls, we go to <u>Round 2</u>. This is a 2 x 2 game played simultaneously (each player being unable to communicate with the other). The payoff matrix is

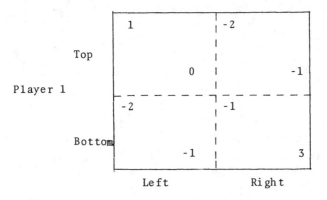

Player 2

a) Write the normal form of the game and compute the sophisticated equilibrium. Notice that Player 1 wins the competition for the first move of the 2 x 2 game. Interpretation? b) Consider now the <u>new</u> game where in Round 0 Player 2 has the opportunity to fold and guarantee the payoffs (0, 1/2). Write the normal form of the game and compute the sophisticated equilibrium. c) You are Player 1 involved in the game of question b and Player 2 <u>does</u> call. In what sense do you think this is a rational move of your opponent? What do you play?

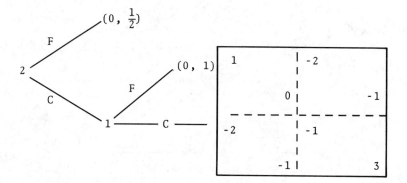

Solution:

(a) The sophisticated equilibrium is x_1 = (Call, Top), x_2 = Left. Player 1 can guarantee zero by Folding; so, if he calls, it must be to get a better payoff, which is only possible by winning the competition for the first move in the 2 x 2 game. The normal form of the game is:

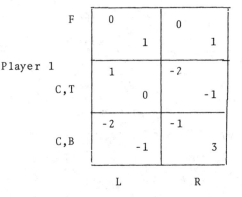

F dominates (C, B) for Player 1; next L dominates R for Player 2; next (C, T) dominates F.

(b) The new matrix is

F	0		0		0	
		1		1		1/2
	1		-2		0	
		0		-1		1/2
	-2		-1		0	
		-1		3		1/2

F dominates (C, B) for Player 1; next (C, L) dominates (C, R) for Player 2; next (C, T) dominates F for Player 1; next F dominates (C, L) for Player 2.

The sophisticated equilibrium is ((C, T), F) with payoff (0, 1/2).

(c) If Player 2 Calls, Player 1 either folds (prudent behavior) or play Call-Top (sophisticated behavior). Player 2 expects that 1 will Fold, so he (Player 2) receives +1. So, Call-Top is a better answer for Player 1 only if he believes that Player 2 was bluffing while calling; however if Player 1

believes in Player 2's implicit threat to play Right if
Player 1 calls, he (Player 1) should better Fold and make
Player 2's threat a success.

2) Fi rst-price auction

Consider the firstprice auction game as defined in
Example 1, Chapter 3. a) Perform the successive elimination
of dominated strategies (two rounds are enough). Notice that
the tie-breaking rule makes the computation of X^1_i slightly
different when i = 1 or i ≥ 2. b) Prove that our game is not
dominance solvable, nor is the reduced game inessential.
However, in the reduced game, prudent behavior is in general
deterministic. Hence, a natural noncooperative outcome of
the game results.

Solution:

(a) First, $D_1(u_1, X) = X^1_1 = [0, a_1[$. Namely, take
any x_1, $x_1 ≥ a_1$, and y_1, $y_1 < a_1$. Then

for all x_{-1}: $u_1(x_1, x_{-1}) ≤ 0 ≤ u_1(y_1, x_{-1})$

for some x_{-1} such that $1 ∈ w(y_1, x_{-1})$: $0 < u_1(y_1, x_{-1})$

Thus y_1 dominates x_1. However, for any two y_1, $z_1 ∈ [0, a_1[$,
none dominates the other. If $y_1 < z_1$, say, and $w(y_1, x_{-1}) ∌$
$1 ∉ w(z_1, x_{-1})$, then z_1 is better than y_1. If $w(y_1, x_{-1}) =$
$w(z_1, x_{-1}) = 1$, then y_1 is better than z_1.

By the same argument for any $i \geq 2$, $\mathcal{D}_i(u_1, X) = X_i^1 =]0, a_i[$, the only difference being that strategy zero for $i \geq 2$ yields always zero hence is dominated by any $y_i \in]0, a_i[$. Second round of elimination: if some Player i* values the object strictly higher than anybody else, $a_i^* > \sup_{i \neq i*} a_j = \alpha$. Then

$$\mathcal{D}_i^*(u_i^*, X^1) = X_{i*}^2 =]0, \alpha] \text{ (or } [0, \alpha] \text{ if } i* = 1)$$

since any strategy $x_{i*} \geq \alpha$ will be the highest bid in X^1. However, for all $i \neq i*$ the same argument as above shows that $X_i^2 = X_i^1$. Thus in this case

$$X_{i*}^\infty =]0, \alpha]$$

and for all $i \neq i*$

$$X_i^\infty =]0, a_i[$$

(For $i = 1$ add zero).

The second case is when at least two players have the highest value. In that case, none of these players can safely lower his bid (no bid in X_i^1 guarantees to get the object). So,

$$X_i^\infty = X_i^1 =]0, a_i[$$

for all i (for $i = 1$ add 0).

(b) In X^∞ no pair x_i, y_i of strategies for Player i are equivalent (see above), so our game is not dominance solvable.

Prudent behavior in game $(X^{\infty}_i, u_i, i = 1, \ldots, n)$. If some Player i* is such that $v_i* > \alpha = \sup_{i \neq x*} v_i$, then Player i*'s unique prudent strategy in X^{α}_i* is α, guaranteeing the profit $v_i* - \alpha$. For any other player all strategies are equally prudent.

If at least two players value the object most $(v_i* = v_j* = \sup_i v_i)$, then all strategies of all players are prudent in X^{∞} (guaranteeing a zero profit only), so in that case prudent behavior in $(X^{\infty}_i, u_i, i = 1, \ldots, n)$ is not deterministic.

3) A game of ascending auction (O'Neill [1985])

A prize of \$n, $n \geq 1$, is auctioned off between two players. Each player has a total wealth of \$r, $r \geq 1$. The players, starting with Player 1, take turns to raise bids, with a minimal increment \$1. Once a player — say i — does not overbid his opponent's last bid, the game is over. Player j gets the prize and <u>both</u> bids are paid.

All variables in this game, n, r as well as all bids, are integer valued. To extract a specific subgame perfect equilibrium we make the following assumption. When the revenue from bidding or not bidding are the same, a player refrains from bidding. a) Show there is a unique subgame perfect equilibrium under the assumption made above and that

Player 1 bids f(r, n), while Player 2 bids nothing (so the game stops after one round). b) Prove that f(r, n) is worth

$$f(r, n) = \text{remainder of the division or } r \text{ by}$$
$$n - 1, \text{ if } r/n - 1 \text{ is not an integer}$$
$$= n - 1, \text{ if } r/n - 1 \text{ is an integer}$$

Solution:

(a) In Kuhn's reduction algorithm, we have room for several subgame perfect equilibria. When the revenue from bidding or not bidding are the same, a player can go either way. The additional assumption selects a unique such equilibrium.

At this equilibrium somebody must win the object and make some profit (otherwise he abstains by the assumption). It must be Player 1, for, in the game starting after Player 1 bids nothing, Player 2 expects the same subgame perfect equilibrium payoff as Player 1 in the original game. So, Player 1 starts by bidding some f(r, n) > 0; next Player 2 bids nothing because in the subgame starting by his move, he expects a zero revenue (since both players cannot make a profit).

(b) We compute now f(r, n) by induction. We know f(r, n) ≤ n - 1 (otherwise 1 makes no profit). Also f(r, n) ≥ 1 (if f(r, n) = 0 Player 1 would refrain from bidding in the first place, so Player 2 could guarantee a

1 - profit by bidding (n - 1) next).

For $r \leq n - 1$, we have $f(r, n) = r$, since an initial bid $x < r$ by Player 1 allows Player 2 to guarantee a positive profit by bidding r.

Suppose now $r \geq n$. After Player 1's opening bid $x = f(r, n)$, if Player 2 decides to bid $y > x$, he would bid (by the subgame perfectness assumption)

$$y = x + f(r - x, n)$$

and make a total profit $n - f(r - x, n) - x$. Thus, he will refrain from bidding iff $y \geq n$. Hence the induction formula

$$f(r, n) = \inf \{x/1 \leq x \leq n - 1 \text{ and } n \leq f(r - x, n) + x\}$$

This leads to the announced formulas for f.

4) Variant of the Steinhaus method

In Example 3 (the Steinhaus method to share a cake) the knife is now moved from $x = 1$ to $x = 0$. How is the previous analysis affected?

Solution:

The prudent strategies are, again, $t_1 = (3 - \sqrt{5}/2)$, $t_2 = (\sqrt{5} - 1/2)$. Now, however, Player 2 stops the knife first and gets $[t_2, 1]$, while Player 1 gets $[0, t_2]$ i.e., the whole surplus. Elimination of dominated strategies results in

48

$$X^1_1 = [0, t_1] \qquad X^1_2 = [0, t_2]$$

$$X^2_1 = X^1_1 \qquad X^2_2 = [0, t_1]$$

If both stop the knife at t_1, the rule is that Player 1 gets $[0, t_1]$ and Player 2 gets $[t_1, 1]$; that is to say, 2 keeps the whole surplus. Thus for Player 2, strategy t_1 dominates any $x_2 > t_1$. If the rule breaks ties in the opposite direction, then Player 2 should stop the knife just before t_1 (although X^2_2 is still $[0, t_1]$!).

5) The divide-and-choose method

The cake and player's tastes are the same as in Example 3, but the method is different. Now Player 1 picks a number $x_1 \in [0, 1]$. Next Player 2 chooses either one of $[0, x_1]$ or $[x_1, 1]$ as his share (and Player 1 gets the remaining share). a) State the normal form of the game. b) Prove that each player has a unique prudent strategy and that prudent behavior allows Player 2 to keep all the surplus. c) Assuming that Player 2, when indifferent about two shares, chooses so as to favor Player 1, prove that the corresponding subgame perfect equilibrium gives all the surplus to Player 1.
d) Compute the optimal utility level of either player acting as a leader. Interpretation?

Solution:

(a) Here $X_1 = [0, 1]$ and $X_2 = \{L, R\}^{[0, 1]}$, i.e., the

set of mappings $X_1 \to \{$Left, Right,$\}$ specifying for each x_1
the choice of Player 2: $[0, x_1]$ or $[x_1, 1]$.
Thus,

$$u_1(x_1, x_2) = v_1([x_1, 1]) \quad \text{if } x_2(x_1) = L$$

$$= v_1([0, x_1]) \quad \text{if } x_2(x_1) = R$$

$$u_2(x_1, x_2) = v_2([0, x_1]) \quad \text{if } x_2(x_1) = L$$

$$= v_2([x_1, 1]) \quad \text{if } x_2(x_1) = R$$

(b` Prudent behavior of Player 1 has him solve

$$\sup_{0 \leq x_1 \leq 1} \inf \{v_1[x_1, 1], v_1([0, x_1])\}$$

Hence his unique prudent strategy is t_1 such that

$$v_1[t_1, 1] = v_1[0, t_1] \quad \text{namely} \quad t_1 = 3 - \sqrt{5}/2$$

The prudent behavior of Player 2 is also his dominating
strategy x_2^* ;

given x_1 choose L if $v_2[0, x_1] > v_2[x_1, 1]$

choose R if $v_2[0, x_1] < v_2[x_1, 1]$

choose R if $v_2[0, x_1] = v_2[x_1, 1] = x_1 = \dfrac{\sqrt{5} - 1}{2}$

(since Player 2 is assumed to be nice to Player 1 if he is
indifferent between the two pieces).

The prudent outcome is $t_1 = (3 - \sqrt{5})/2$, $x_2^*(t_1) = R$. Player 2 keeps all the surplus.

(c) Here $X_1^1 = X_1$ since for any two x_1, y_1, such that $x_1 < y_1$, x_1 is better than y_1 against a leftist Player 2, whereas y is better than x_1 against a rightist Player 2.

Also by (b) (and the special assumption about ties), $X_2^1 = \{x_2^*\}$ (where x_2^* is the prudent dominating strategy described in b). Next $X_1^1 = \{$Best reply of 1 to $x_2^*\}$ = $\{(\sqrt{5} - 1)/2\}$, and Player 1 keeps all the surplus. Without the tie breaking assumption, Player 1 would do almost as well by picking $x_1 = [(\sqrt{5} - 1)/2] - \varepsilon$.

(d) Player 1, acting as a leader, keeps all the surplus by using $x_1^* = (\sqrt{5} - 1)/2$. Player 2 acting as a leader commits himself to x_2:

$$\text{choose L if } v_1([0, x_1]) < v_1([x_1, 1])$$

$$\text{choose R if } v_1([0, x_1]) > v_1([x_1, 1])$$

If $x_1 = (3 - \sqrt{5})/2$, choose R.

In this way he keeps all the surplus.

6) <u>Dividing a shrinking dollar</u> (Rubinstein [1982])

Agents 1, 2, ..., n divide a dollar among themselves according to the following rule:

<u>Step 1</u>: Player 1 proposes a sharing $x^1 = (x_1^1, \ldots, x_n^1)$

where $\sum\limits_{i=1}^{n} x^1_i = 1$ and $x^1_i \geq 0$, all $i \in N$. Then agents 2, ..., n each have the option to accept x^1 or reject it. If all agents agree on x^1, it is done. If at least one agent rejects x^2, then we go to Step 2.

Step 2: Player 2 submits a proposal x^2 to the unanimous approval of the other agents. If his request is rejected, we go to Step 3 where agent 3 makes a proposal, and so on. If step n is ever reached, and Player n's proposal is rejected, then the whole procedure starts again, with a proposal by Player 1, and so on.

We assume that the initial dollar is depreciated at each period by a discount factor τ, $0 < \tau < 1$. Thus, at period 2, it remains $\delta = 1 - \tau$ dollar to be shared, δ^2 at period 3 and so on. Of course in the event that the division procedure goes on undefinitely, each player ultimately makes a zero profit.

Consider the following simple strategies (much simpler than arbitrary strategies can be in this infinitely long game). Each player, say i, has an acceptance level x_i; he accepts at period t any offer from which he would get at least $\delta^{t-1} \cdot x_i$ and rejects any other. In addition, Player i has a fixed proposal strategy $y^i = (y^i_1, ..., y^i_n)$. At time t he proposes the allocation $\delta^{t-1} \cdot y^i$.

Prove that a proper (unique) choice of x^i, y^i, $i = 1, ..., n$ makes the corresponding strategy n-tuple a subgame perfect equilibrium of our game. The proof that all subgame perfect equilibria of this game (using even nonsimple strategies) achieve the same sharing of our dollar is more complicated (see Rubinstein [1982]).

Solution:

Let (x_1, y^1), ..., (x_n, y^n) be a n-tuple of simple strategies. Say that proposal y^1 by Player 1 is accepted by players 2, ..., n:

$$y^1_i \geq x_i \quad \text{all} \quad i \geq 2$$

Thus, given x_2, ..., x_n, Player 1 optimally extracts $1 - (x_2 + \ldots + x_n)$ by proposing $y^1_i = x_i$ to every other player. A similar argument for Player 2, ..., Player n gives $y^j_i = x_i$, all $i \neq j$.

Suppose Player 1's proposal in the first period is rejected. Then the game starting by Player 2's proposal is the same as the original game by accurately permuting the players and discounting the dollar by δ. Hence

$$x_2 = \delta x_1, \; x_3 = \delta x_2, \; \ldots, \; x_n = \delta x_{n-1}$$

Moreover, equality $x_i + \sum_{j \neq i} y^i_j = 1$ follows from the equilibrium property. Thus,

$$x_1 + x_2 + \ldots + x_n = 1$$

The unique solution to the above system is

$$x_i = \frac{\delta^{i-1}}{1 + \delta + \ldots + \delta^{n-1}} \qquad \text{all } i = 1, \ldots, n$$

b) *Abstract games*

7) Sophisticated equilibria generalize dominant strategy equilibria

If game G has (at least) one equilibrium in dominating strategies, then G is dominance solvable and its sophisticated equilibrium coincide with its dominant strategy equilibrium. Hint: Use Exercise 7, Chapter 3.

Solution:

Since $D_i(u_i)$ is nonempty, by Exercise 7, Chapter 3, we have $\mathcal{D}_i(u_i) = D_i(u_i)$; hence, $X_i = D_i(u_i)$. Moreover, all strategies of $D_i(u_i)$ are equivalent, so that $D_i(u_i) = X_i^2 = X_i^3 = \ldots$, and finally $X_i^\infty = D_i(u_i)$.

8) Stackelberg equilibrium

Let $G = (X_1, X_2, u_1, u_2)$ be a two-player game where X_i, $i = 1, 2$ are finite and u_i, $i = 1, 2$ are one-to-one on $X_1 \times X_2$. To analyze the behavior of both players when Player 1 is a leader (has the first move) we consider the game

$$L(G, 1) = (X_1, X_2^{X_1}, \tilde{u}_1, \tilde{u}_2) \text{ where}$$

$\cdot \zeta_2 \; \varepsilon \; X_2^{X_1}$ is any mapping $x_1 \to \zeta_2(x_1)$ from X_1 into X_2.

$\cdot \tilde{u}_i$, $i = 1, 2$, is defined by

$$\tilde{u}_i(x_1, \; \zeta_2) = u_i(x_1, \; \zeta_2(x_1))$$

a) Comment upon this definition. b) Prove that $L(G, 1)$ is dominance solvable and that its sophisticated equilibrium corresponds to the 1-Stackelberg equilibrium of G (Chapter 2, Section 3). c) Compute Player 2's optimal utility level as a leader in $L(G, 1)$. How does this relate to the Wolf-Sheep Lemma in Chapter 2, Section 4?

Solution:

(a) Player 2 chooses second and can carry out any threat ζ_2, whether credible or not.

(b) Player 2 has a dominating strategy ζ_2^*, namely, the best response mapping (unique by the one to one assumption)

$$u_2(x_1, \; \zeta_2^*(x_1)) = \sup_{x_2} u_2(x_1, \; x_2) \qquad \text{all } x_1 \; \varepsilon \; X_1$$

So, $X_2^1 = \{\zeta_2^*\}$. Thus X_1^2 is the (unique) best reply of Player 1 to ζ_2^*, namely, x_1^*, defined by

$$u_1(x_1^*, \; \zeta_2^*(x_1^*)) = \sup u_1[x_1, \; \zeta_2^*(x_1)]$$

The game $L(G, 1)$ is d solvable with sophisticated eq. $(x_1^*, \; \zeta_2^*)$. At this equilibrium Player 1 acts optimally as a leader and gets the corresponding Stackelberg utility S_1.

(c) If Player 2 is leader in $L(G, 1)$ he can do exactly
as much as in the Wolf-Sheep lemma. Define x^0 by

$$u_1(x^0) \geq \alpha_1, \ u_2(x^0) = \max \ \{u_2(x)/u_1(x) \geq \alpha_1\}$$

By the one-to-one assumption , this outcome is unique.
Then Player 2 gets utility $u_2(x^0)$ by picking, as a <u>first</u>
move in $L(G, 1)$, the following threat ζ_2^0 :

$$\zeta_2(x_1^0) = x_2^0$$

For any $x_1 \neq x_1^0$; $\zeta_2(x_1)$ is as bad as possible for
Player 1 $u_1(x_1, \ \zeta_2(x_1)) = \inf_{x_2} u_1(x_1, \ x_2) \leq \alpha_1.$

9) <u>Games in extensive form and subgame perfection</u>
Prove that every finite game in extensive form has at
least one subgame perfect equilibrium.

Solution:

The point of the exercise is to drop the one-to-one
assumption and still prove the existence of a subgame perfect
equilibrium (although the game might not be dominance solvable).
Proof by induction on the game tree. Suppose the claim
hold for all game trees with length at most p and consider
a game tree of length p + 1. As in Definition 3, let L(M)
be the subset of modes of which all successors are terminal
modes. For each m in L(M) ∩ M_j select an optimal successor

m_j of m w.r.t. utility u_i (as in Definition 3). This allows us to reduce the initial game to a game of length p where by the induction assumption there exists a subgame-perfect equilibrium. To make it a subgame-perfect equilibrium of the original game it is enough to specify that, at $m \in L(M) \cap M_j$, Player j chooses precisely the successor m_j selected above.

CHAPTER 5   NASH EQUILIBRIUM

a   *Games in normal form*

1) <u>Auction dollar game</u>

Consider the two-person game of Example 2, Chapter 2.
Compute the best reply correspondences and the Nash equilibria. Are they Pareto-optimal outcomes?

Solution:

The game is symmetrical

$$BR_1(x_2) = \begin{array}{ll} 0 & \text{if } x_2 > 1 \\ 0 \cup \,]1, +\infty[& \text{if } x_2 = 1 \\ ]x_2, +\infty[& \text{if } x_2 < 1 \end{array}$$

The Nash equilibria are

$$\{0\} \times]1, + \infty[\quad \text{and} \quad]1, + \infty[\times \{0\}$$

one player overbids and the other does not bid at all.

They are Pareto efficient (one player keeps all the surplus). There is a third Pareto efficient (and fair) outcome, namely, $(0, 0)$ with payoffs $(1/2, 1/2)$.

2) First-price auction

Consider the first-price auction game (defined in Example 1, Chapter 3; see also Exercise 2, Chapter 4) and suppose the values of the object to the players are such that

$$a_1 > a_2 \geq a_j \qquad \text{all } j \geq 2$$

Prove that the Nash equilibria are such that Player 1 gets the object at some price p, $a_2 \leq p \leq a_1$. Prove that the range of p is the whole interval $[a_2, a_1]$.

Solution:

Let x^* be a Nash equilibrium where Player i, $i \geq 2$ gets the object. Then $x^*_i \leq a_i$. Otherwise, i suffers a loss $(u_i(x^*) = a_i - x^*_i < 0)$ and would be better off by bidding $x_i = 0$. Since i gets the object, $u_1(x^*) = 0$. Yet $x_1 = (a_1 + a_2)/2$ would make Player 1 the winner of the auction (since $x_1 > a_2 \geq a_i \geq x_i$) with a profit:

$$u_1(x_1, x^*_{-1}) = a_1 - x_1 = \frac{a_1 - a_2}{2} > 0 = u_1(x^*)$$

Contradiction.

Thus at a Nash equilibrium, Player 1 gets the object. Pick a Nash equilibrium $\tilde{x}$: as above $\tilde{x}_1 \leq a_1$. Suppose next $\tilde{x}_1 < a_2$. Then by bidding $x_2 = (a_2 + \tilde{x}_1)/2$ Player 2 would win the auction and make a profit. Thus $a_2 \leq \tilde{x}_1 \leq a_1$.

Conversely given p, $a_2 \leq p \leq a_1$, then any $\tilde{x}$ where $\tilde{x}_1 = p = \tilde{x}_2 \geq \tilde{x}_i$ i = 3, 4, ..., n, is a Nash equilibrium. Notice that Player 2 uses a dominated strategy $\tilde{x}_2$ (above his value a_2). Otherwise Player 1 could get the object cheaper.

3) Second-price auction

In the second-price auction game (Example 1, Chapter 4) show that Nash equilibria are very numerous. More precisely, for any Player i and any price p, such that $0 < p \leq a_i$, prove the existence of at least one Nash equilibrium where Player i gets the object and pays p for it.

Solution:

Fix i and p, $0 < p \leq a_i$ and set $x^*_i = \sup_{1 \leq j \leq n} a_j + 1$; $x^*_j = p$ all $j \neq i$. Then x^* is a Nash equilibrium. Player i's only option is to let someone else get the object, in which case he makes a zero profit; Player j's, $j \neq i$, only option is to get the object at price x^*_i, hence above a_j.

4) Another counterintuitive noninessential game

Give an example of a 2 x 2 game (two players, two strategies each) where each player has a unique prudent strategy x^*_i ; where $x^* = (x^*_1 , x^*_2)$ is a Nash equilibrium and a Pareto-optimal outcome; and x^* is the only outcome such that $\alpha_i \leq u_i(x^*)$ $i = 1, 2$; yet the game is not inessential since $\alpha_i < u_i(x^*)$ for $i = 1, 2$.

Solution:

$$\begin{array}{c|c|c|}
\multicolumn{1}{c}{} & \multicolumn{1}{c}{} \\
x^*_1 & 2 \qquad\quad & 1 \qquad\quad \\
 & \quad\qquad 2 & \quad\qquad 1 \\
\hline
 & 0 \qquad\quad & 3 \qquad\quad \\
 & \quad\qquad 1 & \quad\qquad 0 \\
\hline
\end{array}$$

$$x^*_2$$

5) <u>On multiple Nash equilibria and competition for the first move</u>

Suppose the two-person game $G = (X_1, X_2, u_1, u_2)$ has two Pareto-optimal Nash equilibria x, y, with distinct utility vectors $(u_1(x), u_2(x)) \neq (u_1(y), u_2(y))$.

Prove that the competition for the first move arises in G.

Solution:

Denote $(u_1, u_2)(x) = (a_1, a_2)$ and $(u_1, u_2)(y) = (b_1, b_2)$. By Pareto optimality of x and y respectively, $a_1 < b_1 \Rightarrow a_2 > b_2$ and $a_1 > b_1 \Rightarrow b_2 > a_2$. Without loss, take

$a_1 < b_1$, $b_2 < a_2$. Since x is a Nash equilibrium, $S_1 \geq b_1$; since y is a Nash equilibrium, $S_2 \geq a_2$; hence, (S_1, S_2) is Pareto superior to (a_1, a_2) (as well as to (b_1, b_2)). Thus (S_1, S_2) is not a feasible utility vector (otherwise we contradict Pareto optimality of x).

b) *Games in extensive form*

6) <u>A three step competition for the first move</u>

Consider the following game

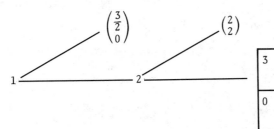

Write the normal form of the game. Compute its sophisticated equilibrium x*. Prove that our game has one more Nash equilibrium y* and that y* Pareto dominates x*. Explain how the players can enforce y*.

Solution:

		T	L	R
	T	3/2 0	3/2 0	3/2 0
Player 1	M	2 2	3 1	0 0
	B	2 2	0 0	1 3
		T	L	R

For Player 2 L is dominated (by T). Next for Player 1, M is dominated (by B). Next for Player 2, T is dominated by R, so the sophisticated equilibrium is $x^* = (T, R)$ with payoffs $(3/2, 0)$. The other Nash equilibrium is $y^* = (M, T)$ with payoffs $(2, 2)$. To enforce this equilibrium Player 2 must promise Player 1 to refrain from calling the 2 x 2 game if Player 1 gives him the move.

7) Voting by veto

Consider the game of voting by veto described in Example 4, Chapter 4. Which candidates can be elected at some Nash equilibrium? Answer the same question in the new game where Player 1's preferences are now

$$u'_1 (c) < u'_1 (b) < u'_1 (d) < u'_1 (a)$$

Solution:

The Nash equilibrium candidates are a and b. The outcome a results from the sophisticated equilibrium. Here is a Nash equilibrium with outcome b:

x^*_1 : eliminate d

x^*_2 : eliminate c if x_1 = a or d
eliminate a if x_1 = b or c

x^*_3 : dominating strategy of Player 3 (i.e., eliminate whichever outcome is worst to u_3 among the two survivors)

Clearly $\pi(x^*_1, x^*_2, x^*_3) = b$. Also Player 3 has no incentive to deviate since x^*_3 is dominating.

Next, Player 2 has no incentive to deviate, since b is his best preferred outcome. Finally, Player 1 has no incentive to deviate, since, given x^*_2 and x^*_3, any strategy $x_1 \neq x^*_1$ would force the election of d.

For the new preference of Player 1 the Nash equilibrium candidates are a (sophisticated), and b, d.

c) *Oligopoly models*

8) Cournot duopoly with fixed costs

This is a quantity-setting oligopoly where both firms supply respectively the quantities x_1, x_2 and the resulting price (at which all supply is sold) is $p = 1 - x_1 - x_2$. Production involves a fixed cost $a > 0$ to produce any positive amount of the good, and no variable cost. Hence the game

$$X_1 = X_2 = [0, 1]$$

$$u_i(x_1, x_2) = x_i(1 - x_1 - x_2) - a \qquad \text{if } x_i > 0$$

$$= 0 \qquad\qquad\qquad \text{if } x_i = 0$$

We discuss this game with respect to the parameter a.
a) Show that for a small ($0 < a < \alpha$ for some α to be computed) the game is strategically similar to the costless game ($a = 0$).

There is a unique Nash equilibrium; it is also the sophisticated equilibrium of the game. b) For a not-too-small and not-too-big ($\alpha \leq a \leq \beta$ for some β to be computed) there are three equilibria (in pure strategies). One of them is the sophisticated equilibrium. c) For a large ($\beta < a \leq 1/4$) there are two asymmetrical equilibria in pure strategies, where only one firm is active. In this case there is also a symmetrical equilibrium in mixed strategies.

Solution:

(a) $0 < a < (1/16)$. The best replies are

$$Br_1(x_2) = \frac{1}{2}(1 - x_2) \qquad \text{if } x_2 \leq z$$

$$= 0 \qquad \text{if } x_2 \geq z$$

where $z = 1 - 2\sqrt{a} > (1/2)$

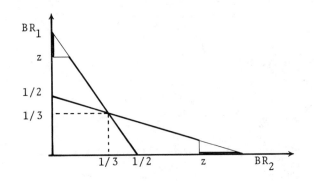

Here $(1/3, 1/3)$ is the unique NE and the sophisticated equilibrium as well.

(b) $\frac{1}{16} \leq a \leq \frac{1}{9}$

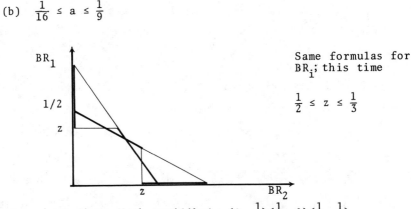

Same formulas for BR_i; this time

$$\frac{1}{2} \leq z \leq \frac{1}{3}$$

So that we have three Nash equilibria $(0, \frac{1}{2}) (\frac{1}{2}, 0) (\frac{1}{3}, \frac{1}{3})$ and $(1/3, 1/3)$ is the sophisticated one.

(c) $\frac{1}{9} < a \leq \frac{1}{4}$

Only $(0, 1/2)$ and $(1/2, 0)$ are left, as NE in pure strategies.

9) <u>Duopoly à la Bertrand</u>

Two firms sell the same good. Their strategy is to establish a <u>price</u> $x_i \geq 0$. If $x_i < x_j$, firm i must satisfy <u>all</u> the demand $D(x_i) = 300 - 5x_i$ and firm j sells nothing. If $x_1 = x_2$, the firms share equally the demand $D(x_1)$. a) Assume first that the cost function is $C(q) = 10q$ for both players. Write the normal form of the game, with strategy sets

$$X_1 = X_2 = [0, 60]$$

Show that there is a unique Nash equilibrium: compare the corresponding profits to the minimal guaranteed profits.

b) Assume next that the cost functions differ

$$C_1(q_1) = 10q_1 \quad C_2(q_2) = 20q_2$$

Prove that, strictly speaking, there is no Nash equilibrium in this game. However, define the ε Nash equilibrium as an outcome where no player, by a unilateral deviation, can improve his utility by more than ε. Then prove that there are infinitely many ε Nash equilibria in this game and describe them. c) In the game in b) perform two rounds of elimination of dominated strategies; show that the reduced game is inessential and its equilibrium corresponds to the ε Nash equilibrium most favorable to Player 1.

Solution:

(a) The normal form of the game is expressed as follows.

$$X_1 = X_2 = [0, 60]$$

$$u_1(x_1, x_2) = (300 - 5x_1)(x_1 - 10) \quad \text{if } x_1 < x_2$$

$$\tfrac{1}{2}(300 - 5x_1)(x_1 - 10) \quad \text{if } x_1 = x_2$$

$$0 \qquad\qquad\qquad\qquad \text{if } x_2 < x_1$$

$$u_2(x_1, x_2) = u_1(x_2, x_1)$$

We compute Player 1's best reply mappings

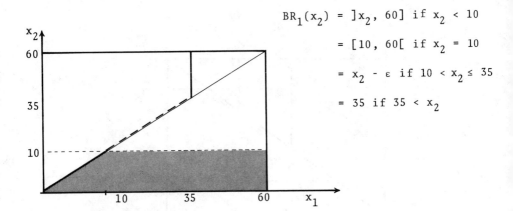

$$BR_1(x_2) =]x_2, 60] \text{ if } x_2 < 10$$
$$= [10, 60[\text{ if } x_2 = 10$$
$$= x_2 - \epsilon \text{ if } 10 < x_2 \leq 35$$
$$= 35 \text{ if } 35 < x_2$$

Thus, formally, $BR_1(x_2)$ is empty if $10 < x_2 \leq 35$ and only then.

Here (10, 10) is the unique Nash equilibrium with payoffs (0, 0) (namely the secure utility level for each player).

(b) The payoff to Player 2 is now

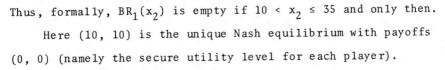

$$u_2(x_1, x_2) = (300 - 5x_2)(x_2 - 20) \qquad \text{if } x_2 < x_1$$
$$= \frac{1}{2}(300 - 5x_2)(x_2 - 20) \qquad \text{if } x_2 = x_1$$
$$= 0 \qquad \text{if } x_1 < x_2$$

His new best-reply function is:

$$BR_2(x_1) =]x_1, 60] \qquad \text{if } x_1 < 20$$
$$= [20, 60] \qquad \text{if } x_1 = 20$$
$$= x_1 - \epsilon \qquad \text{if } 20 < x_1 \leq 40$$
$$= 40 \qquad \text{if } 40 < x_1$$

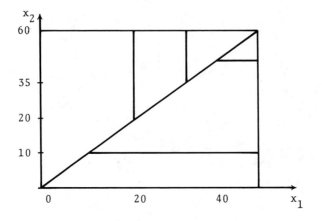

Thus formally there are no NE.

However for all $(x_1, x_1 + \delta)$ where $10 \le x_1 \le 20$ and δ is small, we have an ε Nash equilibrium since Player 1 can gain no more than $\varepsilon = 250\ \delta$ by increasing his price by less than δ, while Player 2 cannot gain from deviating.

(c) The elimination of dominated strategies gives

$$X_1^1 = [10, 35] \qquad X_2^1 = [20, 40]$$

$$X_1^2 = [20 - \varepsilon, 35] \qquad X_2^2 = [20, 35]$$

$$X_i^t = X_i^2 \qquad \text{all } t \ge 3$$

Hence after two rounds of elimination we are left with an <u>inessential</u> game where Player 1 secures (up to ε) profit

$$20(300 - 5x20) - 10.20 = 3800$$

10) Existence of a Nash equilibrium in a quantity-setting duopoly (Fraysse)

The inverse demand function is continuous over $[0, +\infty[$, strictly decreasing and concave on $[0, S]$, and zero after S. The two cost functions c_i, $i = 1, 2$ are continuous on $[0, +\infty]$ and such that:

$$c_i(0) = 0$$

$$c_i(x_i) \geq 1 \qquad \text{for } x_i > 0$$

a) Compute the guaranteed utility level to each player.

b) Show that the best-reply correspondences BR_i are nonincreasing in the following sense:

$$\{x_j < y_j,\ x_i \in BR_i(x_j),\ y_i \in BR_i(y_j)\} \implies \{y_i \leq x_i\}$$

c) Deduce the existence of a Nash equilibrium by picking two single-valued selections b_1, b_2 of BR_1, BR_2 respectively and showing that $b_1 \circ b_2$ has a fixed point on $[0, S]$.

Solution:

$$u_i(x_1, x_2) = x_i p(x_1 + x_2) - c(x_i)$$

(a) For $x_2 = S$ we have $u_1(x_1, x_2) \leq 0$; hence $\alpha_1 \leq 0$. However, $u_1(0, x_2) = 0$ (all x_2); hence $\alpha_1 = 0$.

(b) Take x_1, x_2, y_1, y_2, as in the premises of the implication and write

$$u_i(x_i, x_j) \geq u_i(y_i, x_j)$$

$$u_i(y_i, y_j) \geq u_i(x_i, y_j)$$

Developing and summing up gives

$$x_i p(x) + y_i p(y) \geq y_i p(y_i + x_j) + x_i p(x_i + y_j)$$

Thus

$$x_i[p(x_i + x_j) - p(x_i + y_j)] \geq y_i[p(y_i + x_j) - p(y_i + y_j)]$$

Consider, for fixed $\delta > 0$, the function $\tau \rightarrow \psi(t) = p(t) - p(t + \delta)$

Since p is concave, ψ is nondecreasing in t (this is easily checked if p is twice differentiable; it holds also for any concave function). Thus $x_i < y_i$ would give (take $\delta = y_j - x_j$)

$$0 < \psi(x_i + x_j) \leq \psi(y_i + x_j)$$

As $0 < x_i < y_i$, this implies $x_i\psi(x_i + x_j) < y_i\psi(y_i + x_j)$. Contradiction.

(c) From question b, b_1 and b_2 are nonincreasing; hence, $b_1 \circ b_2$ is nondecreasing from $[0, 1]$ into itself. Claim: a nondecreasing function h from $[0, 1]$ into itself has a fixed point.

$$\exists\ x_1: \quad b_1(b_2(x_1)) = x_1$$

This implies that $(x_1, b_2(x_1))$ is a Nash equilibrium.

Proof of Claim

$$\text{Set } B = \{t \ \epsilon [0, \ 1[\ / h(t) > t\}$$

If B is empty, then $h(0) = 0$. Set $\alpha = \sup \{B\}$ and prove $h(\alpha) = \alpha$. If $h(\alpha) > \alpha$ then $[\alpha, h(\alpha)[\subset B$ (since $\alpha \leq t < h(\alpha) \Rightarrow h(t) \geq h(\alpha) > t$), thus contradicting the definition of α. If $h(\alpha) < \alpha$, then we pick $t \ \epsilon \ B$, $h(\alpha) < t < \alpha$, and, by non-decreasingness of h, $h(t) \leq h(\alpha) < t$. Contradiction.

CHAPTER 6   STABILITY OF NASH EQUILIBRIA

1)   Binary choices with externalities

Analyze the game of Example 1, in the following cases:

a) $a(t) = t$; $b(t) = t + 1/2$; $0 \leq t \leq 1$.   b) $a(t) = (t - 1/3)^2$

$b(t) = (t - 2/3)^2$; $0 \leq t \leq 1$.   In both cases, determine the

noncooperative equilibria of the game (dominant strategies,

sophisticated, Nash stable, and Nash unstable) and find if

they are Pareto-optimal.

Solution:

(a)   This is a prisoner's dilemma type of game: strategy

0 (strictly) dominates strategy 1 so the unique Nash equilibrium

($t = 0$) is also the sophisticated equilibrium; it is stable

and Pareto dominated.

(b)

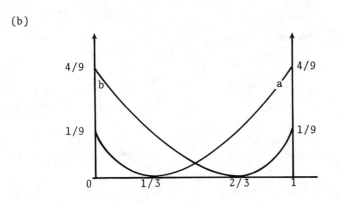

Here we have three Nash equilibria, namely $t = 0$, $t = 1$, and $t = 1/2$. Both $t = 0$ and $t = 1$ are Pareto-optimal and locally stable. In contrast, $t = 1/2$ is locally unstable and Pareto dominated.

2) <u>Quantity-setting oligopoly with limited entry</u>

Ten identical firms produce the same good with the following cost function.

$$c(x_i) = 9x_i + x_i(x_i - 100)^2$$

The inverse demand function is

$$p = 50 - \frac{1}{10} x \qquad \text{if } x \leq 500$$

$$= 0 \qquad \text{for } x \geq 500$$

a) Write the normal form of the game (with unlimited production capacity). Check that the Nash theorem does <u>not</u> apply. b) Prove that in any Nash equilibrium at most four firms can be active (i.e., have $x_i > 0$), then compute all the Nash equilibria of the game. <u>Hint</u>: at any Nash equilibrium the price cannot be smaller than the average cost. c) Are the Nash equilibria of the game locally stable?

Solution:

(a)   Set $\bar{x} = x_1 + \ldots + x_n$. The normal form is

$$u_i(x) \;=\; x_i[41 - \tfrac{1}{10}\bar{x} - (x_i - 100)^2] \quad \text{if } \bar{x} \leq 500$$

$$= -x_i[9 + (x_i - 100)^2] \qquad\qquad \text{if } \bar{x} \geq 500$$

For fixed x_{-i}, u_i is not concave not even quasi concave in x_i.

(b)   If x* is a Nash equilibrium and $x_i^* > 0$ (firm i is "active"), then

$$u_i(x^*) \;\geq\; u_i(0, x_{-i}^*) \;=\; 0$$

Hence, $\bar{x}^* < 500$. (Otherwise, by definition of u_i, $u_i(x^*) < 0$). Moreover,

$$41 - \tfrac{1}{10}\bar{x}^* - (x_i^* - 100)^2 \geq 0 \text{ for all } i \text{ s.t.}$$
$$x_i^* > 0 \qquad (1)$$

Suppose only five firms, say i = 1, 2, ..., 5 are active. Add up these five inequalities:

$$205 \;\geq\; \tfrac{1}{2}\bar{x}^* + \sum_{i=1}^{5} (x_i^* - 100)^2$$

Since $\bar{x}^* = \sum_{i=1}^{5} x^*_i$, this gives:

$$\sum_{i=1}^{5} \{ \underbrace{x^2_i - (200 - \tfrac{1}{2})x_i}_{\lambda} + \underbrace{(10^4 - 4)}_{\mu} \} \leq 0$$

Thus $\lambda^2 - 4\mu \geq 0$, which is not true. The same argument holds if six firms or more are active.

Suppose firms 1, 2, 3, 4 are s.t. $x_i \geq 0$ and $x_5 = \ldots = x_{10} = 0$. The first-order conditions for Nash give

$$i = 1, \ldots, 4 \quad \frac{\partial u_i}{\partial x_i} = 0 \iff 41 - \frac{1}{10}\bar{x} = (x_i - 100)^2 + 2x_i(x_i - 100) + \frac{x_i}{10} \quad (2)$$

Moreover we must have $(\partial^2 u_i / \partial x^2_i) \leq 0$. Hence,

$$6x_i \geq 400 - \frac{1}{5} \Rightarrow x_i \geq 66 \text{ (roughly)}$$

Add now the four inequalities (1):

$$\sum_{i=1}^{4} \{ \underbrace{x^2_i - (200 - \tfrac{2}{5})x_i}_{\lambda} + \underbrace{(10^4 - 41)}_{\mu} \} \leq 0$$

Thus at least one x_i is between the roots of $x^2 - \lambda x + \mu = 0$ which are 99, 16 and 100, 84. So at least one of the x_i is nearly 100.

Back to equation (2) we set $x_i = 100 + y_i$ and we have all y_i solutions of

$$3Y^2 + (200 + \frac{1}{10})Y = C \quad (3)$$

where $C = 31 - (1/10)\bar{x}$. Since the two roots add up to -66

and one of them is in $[-1, +1]$ (corresponding to $x_i \simeq 100$)
the other one is nearly -66. Yet we know $x_i \geq 66 \Rightarrow y_i \geq -44$.
Thus all y_i, $i = 1, 2, \ldots, 4$ are the top root of (3) so
$y_1 = \ldots = y_4$! Thus $\bar{x} = 4x_i$ so equation (2) becomes finally

$$3Y^2 + (200 + \frac{1}{2})Y + 9 = 0$$

whence $Y \simeq -0, 045$ i.e., $x_i \simeq 100 - 0, 045$.

   (c) Compute

$$\frac{\partial^2 u_i}{\partial x_1^2} = -6x_i + (400 - \frac{1}{5}) \simeq -200, \quad \frac{\partial^2 u_i}{\partial x_i \partial x_j} = -\frac{1}{10}$$

   So $t_{ij} \simeq -0, 0005$; thus, the eigen values of T are
$-0, 0005$ and $-0, 00025$. They are much smaller than 1 so
all equilibria are locally stable.

   The clue here is to restrict ourselves locally to the
four players game involving the four active players: the
other one makes a negative profit when they enter. Thus,
their best-reply is, locally, $x_j = 0$.

3) The auto dealer game (Case [1979])

   The n players are n auto dealers facing a constant overall
fixed demand D. Let the strategy x_i be the number of cars
that Dealer i keeps on hand. Assuming that each dealer has
the same number of visitors per unit of time, Dealer i faces
the demand flow

$$D \cdot \frac{x_i}{\overline{x}} \qquad \text{where } \overline{x} = x_1 + \ldots + x_n$$

Let P_i be his unit profit and C_i his unit cost of storing (per unit of time). Then the following normal form game emerges.

$$X_i = [0, +\infty[, \quad u_i(x) = D \cdot P_i \cdot \frac{x_i}{\overline{x}} - C_i x_i \qquad \text{if } \overline{x} > 0$$

$$= 0 \qquad \text{if } \overline{x} = 0$$

a) Prove that every strategy that is large enough is dominated for Player i. b) Fixing the set $I \subset \{1, \ldots, n\}$ of active players (those with $x_i > 0$) write the system characterizing a possible Nash equilibrium. Then give the conditions on the parameters which guarantee the feasibility of that system. c) Study the local stability of the equilibria found in b).

Solution:

Up to a multiplicative constant, the payoffs write

$$u_i(x) = \lambda_i \frac{x_i}{\overline{x}} - x_i, \quad \lambda_i = \frac{DP_i}{C_i} > 0$$

For fixed x_{-i} the payoff to Player i depends only upon $\overline{x}_{-i} = \sum_{j \neq i} x_j$. The function $t \to u_i(t, x_{-i})$ behaves as follows.

For $\overline{x}_{-i} \leq \lambda_i$ it is decreasing on $[0, +\infty[$.

For $0 < \overline{x}_{-i} \leq \lambda$ it increases up to $t = \sqrt{\lambda_i \overline{x}_{-i}} - \overline{x}_{-i}$ then decreases.

For $\overline{x}_{-i} = 0$ it is $\quad \lambda_i - t \qquad \qquad$ if $t > 0$

$$0 \qquad \qquad \text{if } t = 0$$

Best reply of i:

$$\bar{x}_{-i} \geq \lambda_i \rightarrow r_i(\bar{x}_{-i}) = 0$$

$$0 < \bar{x}_{-i} \leq \lambda_i \rightarrow r_i(x_{-i}) = \sqrt{\lambda_i \bar{x}_{-i}} - \bar{x}_{-i}$$

$$\bar{x}_{-i} = 0 \rightarrow r_i(\bar{x}_{-i}) = \varepsilon > 0$$

(a) Observe that

$$\sup_{s \geq 0} \sqrt{\lambda_i s} - s = \frac{\lambda_i}{4}$$

reached for $s = (\lambda_i/4)$; hence $0 < s \leq \lambda_i$. Thus, for all $\bar{x}_{-i}$, the function $t \rightarrow u_i(t, x_{-i})$ is decreasing on $[(\lambda_i/4), +\infty[$. Thus,

$$\mathcal{D}_i(u_i, X) \subset [0, \frac{\lambda_i}{4}]$$

In fact we have equality since any $t \in [0, (\lambda_i/4)]$ is the unique best reply of i for some i.

(b) Let x* be a Nash equilibrium. If all firms are active, $\bar{x}^*_{-i} < \lambda_i$ for all i. Then write $x^*_i = br_i(x^*_{-i})$:

$$x^* = \sqrt{\lambda_i \bar{x}^*_{-i}} \quad \text{all i}$$

$$\Rightarrow \bar{x}^*_{-i} = \frac{(\bar{x}^*)^2}{\lambda_i} \quad \text{all i} \Rightarrow (n-1)\bar{x}^* = \frac{(\bar{x}^*)^2}{\lambda} \quad \text{where}$$

$$\frac{1}{\lambda} = \frac{1}{\lambda_1} + \ldots + \frac{1}{\lambda_m}$$

$$\Rightarrow x^*_i = (n-1)\lambda \left(1 - \frac{(n-1)\lambda}{\lambda_i}\right)$$

This is a feasible strategy only if $(1/\lambda) \geq (n - 1)/\lambda_i$ for all i. However, the initial assumption $\overline{x}^*_{-i} \leq \lambda_i$ is equivalent to $(n - 1)^2 (\lambda)^2 / \lambda_i \leq \lambda_i$ which is the same inequality. Thus a Nash equilibrium where all firms are active exist iff

$$\sum_{j=1}^{n} \frac{1}{\lambda_j} = \frac{1}{\lambda} \geq \frac{n - 1}{\lambda_i} \qquad \text{all } i = 1, \ldots, n$$

This says that the λ_i are close to each other. The characterization of all Nash equilibria is analogous. If the active firms at Nash equilibrium x* are those of $I \subset \{1, \ldots, n\}$,

$$x^*_i > 0 <=> i \in I$$

then define $\lambda(I)$ by $\sum_{j \in I} \frac{1}{\lambda_j} = \frac{1}{\lambda(I)}$. The following system of inequalities must be true.

$$\frac{1}{\lambda(I)} \geq \frac{m - 1}{\lambda_i} \qquad \text{all } i \in I \text{ where } m = |I|$$

$$(m - 1)\lambda(I) \geq \lambda_j \qquad \text{all } j \notin I$$

The second inequality guarantees $\overline{x}^*_{-j} \geq \lambda_j$ for all $j \notin I$. If this holds true, then x* is given by

$$x^*_i = (m - 1)\lambda(I)[1 - \frac{(m - 1)\lambda(I)}{\lambda_i}] \qquad \text{all } i \in I$$

$$x^*_j = 0 \qquad \text{all } j \notin I$$

(c)   Say that x is a Nash equilibrium where all firms are active (the case of an arbitrary I is similar).   Then compute

$$\frac{\partial u_i}{\partial x_i}(x) = \lambda_i \frac{\overline{x}_{-i}}{\overline{x}^2} - 1 \Rightarrow \frac{\partial^2 u_i}{\partial x_i^2}(x) = -2\lambda_i \frac{\overline{x}_{-i}}{\overline{x}^3} = -\frac{2}{(n-1)\lambda}$$

$$\frac{\partial^2 u_i}{\partial x_i \partial x_j}(x) = \frac{\lambda_i}{\overline{x}^2} - 2\lambda_i \frac{\overline{x}_{-i}}{\overline{x}^3} = \frac{\lambda_i}{(n-1)^2\lambda^2} - \frac{2}{(n-1)\lambda}$$

Hence the entry t_{ij} in matrix T is

$$t_{ij} = -t_i = -\left[\frac{\lambda_i}{2(n-1)\lambda} - 1\right]$$

By the same argument as in Example 4 the eigenvalues a of T are all in $]-1, +1[$ if:

$$\sum_{i=1}^{n} \frac{\lambda_i - 2(n-1)\lambda}{\lambda_i} < 1 \iff n \geq 2$$

Hence all Nash equilibria are locally stable.

4)   Stability by sequential Cournot tatonnement

Given an ordering $\{1, 2, \ldots, n\}$ of the players, the $\{1, \ldots, n\}$ sequential Cournot tatonnement starting at x^0 is the sequence $x^0, x^1, \ldots, x^t, \ldots$ where

$$x_i^t = r_i(x_1^t, \ldots, x_{i-1}^t, x_{i+1}^{t-1}, \ldots, x_n^{t-1})$$

if $t = i$ modulo n and $x_i^t = x_i^{t-1}$ otherwise.

Say that a Nash equilibrium x* is {1, ..., n} stable if for any initial position x^0 the {1, ..., n} sequential Cournot tatonnement starting at x^0 converges to x*. a) If n = 2, the {1, 2} stability, the {2, 1} stability, and the stability in the sense of Definition 1 all coincide. b) If n ≥ 3, these notions differ. Consider, for instance, the three players game

$$X_i = R, \ i = 1, 2, 3 \qquad \begin{aligned} u_1(x) &= -(x_1 - x_2)^2 \\ u_2(x) &= -(x_2 - \frac{1}{3} x_3)^2 \\ u_3(x) &= -(4x_1 - 3x_2 - x_3)^2 \end{aligned}$$

The unique Nash equilibrium is (0, 0, 0). Prove that it is not {1, 2, 3} stable, whereas it is {2, 1, 3} stable. Is it stable in the sense of Definition 1?

Solution:

(a) Construct the {1, 2} - sequential Cournot tatonnement starting at x_1^0, x_2^0 :

$$x_1^1 = r_1(x_2^0), \ \ x_2^1 = x_2^0$$
$$x_1^2 = x_1^1, \ \ x_2^2 = r_2(r_1(x_2^0))$$
$$x_1^3 = r_1(r_2(r_1(x_2^0))), \ \ x_2^3 = x_2^2$$

and so on.

The {1, 2} stability of x* means that for any initial x_2 the sequence x_1^{2t+1} converges to x_1^* when t goes to infinity and the sequence x_2^{2t} converges to x_1^* .

Consider now any $x_1 \in X_1$ and set $x_2 = r_2(x_1)$. From {1, 2} stability follows that the sequence

$$r_1[r_2(x_1)], \quad r_1r_2r_1[r_2(x_1)], \quad r_1r_2r_1r_2r_1[r_2(x_1)], \quad \ldots$$

converges to x_1^* and the sequence

$$r_2(x_1), \quad r_2r_1[r_2(x_1)], \quad r_2r_1r_2r_1[r_2(x_1)], \quad \ldots$$

converges to x_2^* . This in turn implies {2, 1} stability. A similar argument establishes that stability in the sense of Definition 1 is equivalent to {1, 2} stability.

(b) The best reply functions are linear.

$$br_1(x_2, x_3) = x_2$$

$$br_2(x_1, x_3) = \tfrac{1}{3}x_3$$

$$br_3(x_1, x_2) = 4x_1 - 3x_2$$

Consider the {1, 2, 3} Cournot tatonnement starting at (x_1, x_2, x_3).

$$x^0 = (x_1, x_2, x_3), \quad x^1 = (x_2, x_2, x_3), \quad x^2 = (x_2, \tfrac{1}{3}x_3, x_3)$$

$$x^3 = (x_2, \tfrac{1}{3}x_3, 4x_2 - x_3), \quad x^4 = (\tfrac{1}{3}x_3, \tfrac{1}{3}x_3, 4x_2 + x_3), \quad \ldots$$

Here $x^3 = Ax^0$ where

$$A = \begin{bmatrix} 0 & 1 & 0 \\ 0 & 0 & \frac{1}{3} \\ 0 & 4 & -1 \end{bmatrix}$$

of which one eigenvalue, $\lambda = -(1/2) - (\sqrt{19/12})$, is greater than 1. Thus, the sequence x^{3t} does not always converge to zero when t goes to infinity.

Consider next the {2, 1, 3} tatonnement:

$$x^0 = (x_1, x_2, x_3), \ x^1 = (x_1, \tfrac{1}{3}x_3, x_3), \ x^2 = (\tfrac{1}{3}x_3, \tfrac{1}{3}x_3, x_3)$$

$$x^3 = (\tfrac{1}{3}x_3, \tfrac{1}{3}x_3, \tfrac{1}{3}x_3), \ x^4 = (\tfrac{1}{3}x_3, \tfrac{1}{9}x_3, \tfrac{1}{3}x_3), \ \ldots$$

Here

$$x^3 = \begin{bmatrix} 0 & 0 & \frac{1}{3} \\ 0 & 0 & \frac{1}{3} \\ 0 & 0 & \frac{1}{3} \end{bmatrix} \cdot x^0$$

so x^{3t} converges to zero when t goes to infinity and so do x^{3t+1}, x^{3t+2}. Finally, consider the simultaneous Cournot tatonnement

$$x^1 = B\, x^0 \quad \text{where } B = \begin{bmatrix} 0 & 1 & 0 \\ 0 & 0 & \frac{1}{3} \\ 4 & -3 & 0 \end{bmatrix}$$

The characteristic polynomial of B is $\lambda^3 + \lambda + (4/3) = 0$ with one real eigenvalue λ_0, $-1 < \lambda_0 < 0$ and two complex eigenvalues with modulus $\sqrt{(4/3|\lambda_0|)} > 1$. Therefore the sequence x^t does not always converge to zero, and our Nash equilibrium is locally unstable.

5) Cournot tatonnement and sophisticated equilibrium
   (Moulin [1984])

a) In the Cournot duopoly with constant return to scales, show that the Nash equilibrium is the sophisticated equilibrium as well.   b) Let G be a two-person game where $X_i = [0, 1]$, u_i is continuous on $[0, 1]^2$ and strictly concave w. r. t. x_i, i = 1, 2.   Thus the best reply of each player is single valued and continuous.   Show that for any subintervals $Y_i \subset [0, 1]$, i = 1, 2 we have

$$D_i(u_i, Y) = \underset{Y_i}{proj}\ r_i(Y_j)$$

Deduce that the successive elimination of dominated strategies is given by

$$x_i^{t+1} = r_i(x_j^t) \qquad\qquad t = 1, 2, ..$$

Conclude that G is dominance solvable iff it has a unique (globally), stable Nash equilibrium.

Solution:

(a) Draw the best reply function for Player 1.

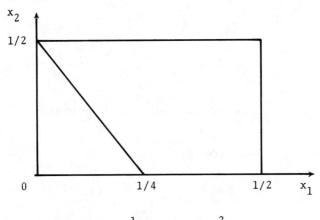

$$u_1(x) = (\tfrac{1}{2} - x_2)x_1 = x_1^2$$

$$\frac{\partial u_1}{\partial x_1}(x) = \tfrac{1}{2} - x_2 - 2x_1$$

For fixed x_2, the function $x_1 \to u_1(x_1, x_2)$ is concave with its peak at $1/4 - (x_2/2)$. Since this peak is $\leq 1/4$ (for all $x_2 \in [0, (1/2)] = X_2$), this means that the function $u_1(\cdot, x_2)$ is always decreasing on $[(1/4), (1/2)]$. Thus, strategies in $](1/4), (1/2)]$ are dominated by $1/4$.

In fact $\mathcal{D}_1(u_1, X) = [0, (1/4)]$ since any strategy $x_1 \in [0, (1/4)]$ is, for some x_2 the unique best reply of Player 1 to x_2. (Hence, it cannot be dominated.)

By the same argument, $\mathcal{D}_2(u_2, X) = [0, (1/4)]$. Hence, the next picture

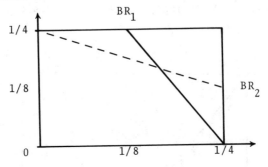

Repeating the above argument we have next

$$x_i^2 = \mathcal{D}_i(u_i, x^1) = [\tfrac{1}{8}, \tfrac{1}{4}]$$

At x^2 the picture is homothetic to the initial one; hence, $x_i^3 = [(1/8), (3/16)]$ and so on... Each step cuts by $1/2$ the strategy set hence an exponential convergence to the Nash equilibrium.

(b) Player i's best reply function r_i is defined over the initial strategy set X_i. If the strategy set shrinks to a subinterval Y_i, the best reply of i over Y_i becomes

$$r_i^{Y_i}(x_j) = \underset{Y_i}{\text{proj}}\ r_i(X_j)$$

This follows from the concavity of u_i (draw a picture). Hence,

$$\underset{Y_i}{\text{proj}}\ r_i(Y_j) = r_i^{Y_i}(Y_j) \subset \mathcal{D}_i(u_i, Y)$$

since any strategy in $r_i^{Y_i}(y_j)$ is the <u>unique</u> best reply to some strategy in Y_j. Conversely, notice that $A_i = \underset{Y_i}{proj}\; r_i(Y_j)$ is a closed interval, since r_i is continuous. If $x_i \notin A_i$, we have for instance the following configuration

$$
\begin{array}{c}
y_i \\
------*------[---------]- \\
\;\;\;\;\;\;\;\; x_i \;\;\;\;\;\;\;\;\;\;\; A_i
\end{array}
$$

Henceforth x_i is dominated by y_i, as the concave function $z_i \rightarrow u_i(z_i,\, x_j)$ is increasing on $[x_i,\, y_i]$ for all x_j (its maximum being reached after y_i). This proves the first equation in b). Then one proves by induction

$$X_i^t \text{ is an interval for all } t = 1,\, 2,\, \ldots$$

$$X_i^t = r_i(X_j^{t-1})$$

Namely we have

$$X_i^{t+1} = \underset{X_i^t}{proj}\; r_i(X_j^t)$$

and, by induction, $r_i(X_j^{t-1}) \subset X_i^t \Rightarrow r_i(X_j^t) \subset X_i^t$

implying $\underset{X_i^t}{proj}\; r_i(X_j^t) = r_i(X_j^t)$

This proves the second equation. If G is dominance solvable, the sequence X_i^t shrinks to a singleton (since no two strategies can be equivalent in X^∞, by strict concavity). Thus the formula $X_i^{t+1} = r_i(X_j^t)$ implies that any Cournot tatonnement sequence converges to that singleton.

CHAPTER 7   MIXED STRATEGIES

a)   *Finite two-person, zero-sum games*

1)   <u>Picking an entry</u>
   Let B $=\begin{pmatrix} 2 & 1 \\ 4 & 5 \end{pmatrix}$.   Player 1 chooses either a row of B or
a column of B.   Player 2 chooses an element of (position in)
B.   If the element picked by Player 2 is in the row or column
chosen by Player 1, then Player 2 pays Player 1 that amount.
Otherwise Player 2 pays nothing.   Solve this game.

Solution:

   The matrix is

	2	1	4	5
Top row	2	1	0	0
Bottom row	0	0	4	5
Left column	2	0	4	0
Right column	0	1	0	5

where the unique mixed equilibrium is

$$\mu_1 = 4/5T + 1/5B \quad \mu_2 = 4/5\{1\} + 1/5 \{4\}$$

with value $x = 4/5$. The clue is to search for a mixed
equilibrium where each player uses two strategies only.

2) One-card, one-round bluffing

With equal probability Player 1 is dealt card H or card L.
Player 2 is not dealt a card, and never gets to look at
Player 1's card until the end of the game. After looking at
his card, Player 1 either plays or folds. If he folds, he
loses $1. If he plays, Player 2 now must consider if he
should fold. If he decides to fold, he loses $1. If he sees,
the card is shown. If it is H, Player 1 wins $4 from Player 2;
if it is L, Player 2 wins $a from Player 2.

Draw a game tree and the normal form of the game. Show
that Player 1 has only two undominated strategies. Find the
mixed value and optimal strategies of this game as a function
of a (restrict yourself to the case a > 1).

Solution:

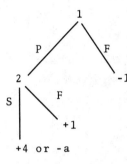

	Sees	Folds	
always P	1/2 (4 - a)	+1	*
always F	-1	-1	
P if H, F if L	3/2	0	
P if L, F if H	-1/2 (a + 1)	0	*

The two star marked strategies are undominated. The
reduced game has no pure saddle point iff a > 2 in which case
the value is (3/a + 1) and optimal strategies are

Player 1: $\dfrac{3}{(a + 1)}$ {always P} + $\dfrac{a - 2}{a + 1}$ {P if H, F if L}

Player 2: $\dfrac{2}{a + 1}$ {sees} + $\dfrac{a - 1}{a + 1}$ {Folds}

If $1 \le a \le 2$, the value is 3/2: Player 2 always Sees and
Player 1 is prudent (P if H, F if L).

3) <u>One-card, two-round bluffing</u>

After looking at his card, Player 1 either plays or folds.
If Player 1 folds, he loses $2. If Player 1 plays, Player 2
now must raise or fold. If Player 2 folds, he loses $2. If
Player 2 raises, then Player 1 must again either play or fold.
If Player folds he loses $4. If Player 1 plays, Player 2 must
now either fold or see. If Player 2 folds, he loses $4. If
Player 2 sees, Player 1 must show the card. If it is H,
Player 1 wins $6; if it is L, Player 1 loses $6. Formally,
Player 1 has nine pure strategies and Player 2 has three (Why?).
Reduce Player 1's strategy set to three strategies by domination
arguments. Then solve the mixed extension of that game.

Solution:

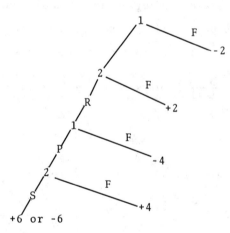

After elimination of dominated strategies, only three
strategies are left for Player 1:

Prudent: Fold immediately if L, always Play if H.

Shy bluff: If L: Play, then Fold if Player 2 Raises;

if H, always Play.

Bluffing: Always Play no matter what.

Player 2 has precisely three strategies.

F: Fold immediately

RF: Raise, next Fold if 1 Plays

RS: Raise, next See if 1 Plays

Hence the matrix form

	F	RF	RS
Pr	0	1	2
SB1	2	0	1
B1	2	4	0

with completely mixed optimal strategies

$$\mu_1 = \begin{bmatrix} 2/5 \\ 2/5 \\ 1/5 \end{bmatrix} \quad \mu_2 = (\frac{5}{15}, \frac{2}{15}, \frac{8}{15}) \quad \text{and value} \quad \frac{6}{5}$$

4) A recursive game

Solve the recursive two-person, zero-sum game:

$$G = \begin{array}{|cc|} \hline 0 & 1 \\ & \\ 1 & -G/2 \\ \hline \end{array}$$

Make up a brief scenario for G.

Solution:

The value v of G is such that

$$v = \text{value} \begin{array}{|cc|} \hline -1 & 1 \\ & \\ 1 & -v/2 \\ \hline \end{array}$$

clearly $-(v/2) < 1$ (otherwise the 2 x 2 game has value 1, a contradition). Thus,

$$v = \frac{(v/2) - 1}{-(v/2) - 3} \iff v^2 + 7v - 2 = 0 \implies v = \frac{\sqrt{57} - 7}{2}$$

A possible story

Player 1 keeps in his hand an odd or even number of pennies. Player 2 tries to guess Player 1's choice. If Player 2 fails, he looses \$1. If he guesses correctly odd, he wins \$1. If he guesses correctly even, the game is played again with the role of both players exchanged and all payoffs divided by half. Strictly speaking, this is an infinite game, but the geometric cutting of the payoffs at each round makes it tractable like a finite game.

5) Another inspector-cheater game

There are n rounds in this game involving a cheater and an inspector. The cheater may cheat in every round. He gains $1 for every round in which he cheats without being inspected. If he is inspected, he is fined $d and is "put out of business." That is, he cannot cheat again. The inspector can inspect only once, and this inspection is "noisy." That is, the cheater knows when it happens. Express this as a recursive game. That is, give G_1 (one round left) explicitly and G_n in terms of G_{n-1}. Check that at round n, the optimal strategy has to be mixed and deduce an induction formula for v_n, the mixed value of G_n, as a function of v_{n-1}. Solve and describe both players' optimal strategies.

Solution:

$$G_1 = \begin{array}{|c|c|} \hline -d & 1 \\ \hline 0 & 0 \\ \hline \end{array}$$
$$\begin{array}{cc} \text{Inspect} & \text{Does not} \\ & \text{inspect} \end{array}$$

with value zero and optimal strategy (Does not cheat, inspect).

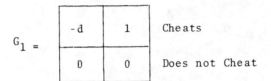

$$G_n = \begin{array}{|c|c|} \hline -d & 1 + G_{n-1} \\ \hline n-1 & G_{n-1} \\ \hline \end{array} \Rightarrow v_n = \text{value of} \begin{array}{|c|c|} \hline -d & 1 + v_{n-1} \\ \hline n-1 & v_{n-1} \\ \hline \end{array}$$

Clearly $v_{n-1} \leq n - 1$ (in face Player 1's payoff can never be above $(n - 1)$) and $v_{n-1} \geq 0$ (by not cheating until inspection has occurred the cheater secures a zero utility). Thus the above 2 x 2 game has no saddle point in pure strategies.

Thus,

$$v_n = \frac{-dv_{n-1} - (n - 1)(1 + v_{n-1})}{-d + v_{n-1} - (n - 1) - (1 + v_{n-1})}$$

$$<=> \quad (n + d)v_n = (n - 1) + (n - 1 + d)v_{n-1}$$

$$=> \quad (n + d)v_n = (n - 1) + (n - 2) + \ldots + 1 = \frac{n(n - 1)}{2}$$

As optimal strategies, the cheater cheats in round n with probability

$$\frac{(n - 1)(n + 2d)}{2(n - 1 + d)(n + d)}$$

which approaches 1/2 for large n. The inspector inspects in round n with probability $1/(n + d)$.

6) Head-tail-head

Each player chooses a triple $\alpha_1 \alpha_2 \alpha_3$ where each α_i is either a head or tail. Thus each has eight different strategies. Then a coin is tossed until one player's figure appears. This player is paid \$1 by the other (of course, if both players choose the same figure, the game is a draw, and no coin has to be tossed).

For instance, take x_1 = TTH, x_2 = THH. If the drawing gives HTHTTH ..., it is a win for Player 1. In fact, given these strategies, the expected payoff to Player 1 is worth 1/3, for after T is drawn and the next one is T, Player 1 wins for sure. However, if the next draw is H, there is a 1/2 probability that Player 2 will win (if the next is H) and a 1/2 probability of landing back at the starting point (if the next draw is T). Show that similar computations for each pair of strategies lead to the accompanying payoff matrix.

	HHH	HHT	HTT	HTH	THT	THH	TTH	TTT
HHH			-1/5	-1/5	-1/6	-3/4	-2/5	
HHT			1/3	1/3	1/4	-1/2		2/5
HTT	1/5	-1/3					1/2	3/4
HTH	1/5	-1/3					-1/4	1/6
THT	1/6	-1/4					-1/3	1/5
THH	3/4	1/2					-1/3	1/5
TTH	2/5		-1/2	1/4	1/3	1/3		
TTT		-2/5	-3/4	-1/6	-1/5	-1/5		

98

Note that blank entries are zero. Also, by symmetry arguments only 11 pair of strategies are nontrivial. Eliminate dominated strategies, then find optimal mixed strategies.

Solution:

The game is symmetrical, so we need only to compute 28 entries. Symmetry arguments reduce this number to 11: e.g. u(HTT, HHH) = u(THH, TTT) and u(HTH, THT) = 0, It remains to compute the payoff at

HHH, HTT	HHT, HTT	HTT, HTH
HHH, HTH	HHT, HTH	HTT, THT
HHH, THT and	HHT, THT and	
HHH, THH	HHT, THH	
HHH, TTH		

To illustrate the method, take HHH, THT. Let α be the probability that Player 1 wins after H is drawn and β the probability that he wins after T is drawn:

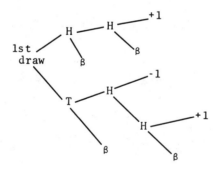

Thus

$$\alpha = \frac{1}{2}\beta + \frac{1}{2}(\frac{1}{2} + \frac{\beta}{2})$$

$$\beta = \frac{1}{2}\beta + \frac{1}{2}(-\frac{1}{2} + \frac{1}{2}(\frac{1}{2} + \frac{\beta}{2}))$$

$$\Bigg\} <=> \Bigg\{$$

$$\alpha = \frac{3}{4}\beta + \frac{1}{4}$$

$$\beta = -\frac{1}{3}$$

with associated payoff

$$\frac{1}{2}\alpha + \frac{1}{2}\beta = -\frac{1}{6}$$

As a next instance take (HHT, THH). With the same definitions of α, β we have

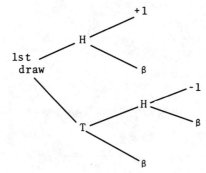

after HH Player 1 wins <u>for</u> <u>sure</u>

$$\alpha = \frac{1}{2} + \frac{\beta}{2}$$

$$\beta = \frac{1}{2}\beta + \frac{1}{2}(-\frac{1}{2} + \frac{\beta}{2})$$

$$\Bigg\} <=> \Bigg\{$$

$$\alpha = 0$$

$$\beta = -1: \text{ after T,}$$
Player 2 wins for sure

We analyze now the matrix game. We have $\sup_{x_1} \inf_{x_2} u = -(1/3) =$ $-(\inf_{x_2} \sup_{x_1} u)$. Moreover strategies HHH, TTT, HTH, and THT are dominated respectively by THH, HTT, HTT, and THH. Hence the reduced game

		1/3	-1/2	
HTT	-1/3			1/2
THH	1/2			-1/3
		-1/2	1/3	

with value zero and optimal strategies HTT + (1 - p)THH for $(2/5) \le p \le (3/5)$.

7) Guessing a number

Player 2 chooses one of the three numbers 1, 2, 5, say, x_2. One of the two numbers not chosen is selected at random and shown to Player 1. Player 1 now guesses which number Player 2 chose, winning $\$x_2$ from him if he is correct. If Player 1 guesses incorrectly the payoffs are zero to both.

Find the mixed value of this game and optimal strategies for both players.

Hint: It is cumbersome to draw the 8 x 3 normal form of this game. Use instead the following description of Player 1's mixed strategies, called "behavioral strategies." If he is shown number 1, he guesses 2 with probability q_1 and 5 with probability $1 - q_1$, etc. Player 1's strategy is now a triple (q_1, q_2, q_5) where each q_k is in $[0, 1]$. Write the corresponding normal form and solve.

Solution:

The normal form of the game with behavioral strategies is as follows.

Player 1: $\left. \begin{array}{l} q_1: \text{ prob. guesses 2 if shown 1} \\ q_2: \text{ prob. guesses 5 is shown 2} \\ q_5: \text{ prob. guesses 1 if shown 5} \end{array} \right\} \quad 0 \le q_i \le 1$

Player 2: $p_1, p_2, p_5 \quad p_1 + p_2 + p_5 = 1 \qquad \text{all } p_i \ge 0$

$$u(p, q) = \tfrac{1}{2}[p_1(q_5 + 1 - q_2) + 2p_2(q_1 + 1 - q_5) + 5p_5(q_2 + 1 - q_1)]$$

$$= \tfrac{1}{2}[q_1(2p_2 - 5p_5) + q_2(5p_5 - p_1) + q_5(p_1 - 2p_2) + 1 + p_2 + 4p_5]$$

Thus strategy sets are convex, and compact, and the payoff is linear in each variable. By von Neumann's theorem, there is a saddle point.

Seeking for a "completely mixed" equilibrium $p_i > 0$, $0 < q_i < 1$ yields

$$p_1 = 2p_2 = 5p_5 \Rightarrow p = (\tfrac{10}{17}, \tfrac{5}{17}, \tfrac{2}{17})$$

$$q_5 + 1 - q_2 = 2(q_1 + 1 - q_5) = 5(q_2 + 1 - q_1) \Rightarrow$$

$$q = (\lambda, \lambda - \tfrac{11}{17}, \lambda + \tfrac{2}{17})$$

where $(1/17) \leq \lambda \leq (15/17)$. The value is $\frac{15}{17}$.

8) Where a completely mixed strategy seems to be optimal

Give an example of a 2 x 2 zero-sum game where Player 1 has a completely mixed strategy μ_1^* such that $u_1(\mu_1^*, \mu_2)$ is independent of $\mu_2 \in M_2$, yet μ_1^* is not optimal in the mixed game. Find a two-person, zero-sum game with no value (in pure strategies) where the same situation arises.

Solution:

0	1
-1	-2

$\mu_1^* = \frac{1}{2}$ Top $+ \frac{1}{2}$ Bottom

1	-1
-1	1
-2	-2

$\mu_1^* = (\frac{1}{3}, \frac{1}{3}, \frac{1}{3})$

Yet the optimal strategy of 1 is $\mu_1 = (\frac{1}{2}, \frac{1}{2}, 0)$ and the value is zero. The trick is to avoid dominated strategies.

9) On games where no optimal strategy is completely mixed

Let $G = (X_1, X_2, u_1)$ be a finite two-person, zero-sum game where no player has an optimal completely mixed strategy.

For all $(x_1, x_2) \in X$ denote $v_1(x_1, x_2)$ the mixed value of the game $(X_1 \setminus \{x_1\}, X_2 \setminus \{x_2\}, u_1)$. Show that the game (X_1, X_2, v_1) has a saddle point in pure strategies and its value is the mixed value of G.

Solution:

Let (μ_1^*, μ_2^*) be a mixed saddle point of G. By assumption, none of μ_1^*, μ_2^* is completely mixed. Thus, for some $(x_1^*, x_2^*) \in X$, we have

$$\mu_1^*(x_1^*) = \mu_2^*(x_2^*) = 0$$

Let v^* be the mixed value of G. Denote by $M_i(x_i)$ the mixed strategies of Player i such that $\mu_i(x_i) = 0$, i.e., the mixed strategies of Player i in the restricted strategy set $X_i \setminus \{x_i\}$. Then we have for all $x_1 \in X_1$

$$v_1(x_1, x_2^*) = \inf_{M_2(x_2^*)} \sup_{M_1(x_1)} \bar{u}_1(\mu_1, \mu_2) \leq \sup_{M_1(x_1)} \bar{u}_1(\mu_1, \mu_2^*) \leq \ldots$$

$$\ldots \leq \sup_{M_1} u_1(\mu_1, \mu_2^*) = v^*$$

A symmetric sequence of inequalities give

$$v^* \leq v_1(x_1^*, x_2) \qquad \text{all } x_2 \in X_2$$

which is the desired conclusion.

10) A property of optimal mixed strategies

If the finite two-person, zero-sum game $G = (X_1, X_2, u_1)$ has a value (in pure strategies), prove that any mixed strategy in G_m, the carrier of which is made solely of optimal strategies of G, is optimal in G_m. Given an example where G_m has more optimal strategies.

Solution:

Let μ^*_1 be a mixed strategy of Player 1, such that

$$\mu^*_1 = \lambda_1 \delta_{x^1_1} + \ldots + \lambda_T \delta_{x^T_1} \quad \text{where } x^t_1 \text{ is optimal in } G \text{ for all}$$

$t = 1, \ldots, T$. Since G has a value v, v is the mixed value as well.

Now we have:

$$\inf_{M_2} \overline{u}(\mu^*_1, \mu_2) = \inf_{X_2} \overline{u}_1(\mu^*_1, \delta_{x_2}) = \inf_{X_2} \{ \sum_{t=1}^{T} \lambda_t u_1(x^t_1, x_2)\} \geq \sum_{t=1}^{T} \lambda_t v = v$$

This proves optimality of μ^*_1.

Here is an example where each player has some optimal mixed strategies which are not convex combinations of optimal pure strategies:

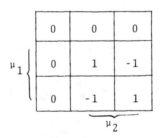

11) <u>The mixed value improves upon both secure utility levels</u>

Let (X_1, X_2, u) be a finite two-person, zero-sum game with no value in pure strategies.

$$\sup_{x_1} \inf_{x_2} u = \alpha_1 < \alpha_2 = \inf_{x_2} \sup_{x_1} u$$

a) Suppose that for each x_1 (respectively for each x_2) the mapping $x_2 \to u(x_1, x_2)$ is one to one on X_2 (respectively $x_1 \to u(x_1, x_2)$ is one to one on X_1). Prove that the mixed value $v_m(u)$ strictly improves the secure utility level of both players

$$\alpha_1 < v_m(u) < \alpha_2$$

b) Give an example where the one to one assumption is false and

$$\alpha_1 = v_m(u) < \alpha_2$$

(<u>Hint</u>: a 3 x 2 game will do).

Solution:

(a) Let x^*_1 be a prudent pure strategy for Player 1 and x^*_2 the best reply to x^*_1 :

$$\alpha_1 = \inf_{x_2} u_1(x^*_1, x_2) = u_1(x^*_1, x^*_2) < u_1(x^*_1, x_2)$$

$$\text{all } x_2 \neq x^*_2$$

In column x_2^* choose $\tilde{x}_1$ such that $u_1(\tilde{x}_1, x_2^*) \geq \alpha_2 > \alpha_1$. Then the mixed strategy $\mu = (1 - \varepsilon)\delta_{x_1^*} + \varepsilon\delta_{\tilde{x}_1}$ guarantees, for $\varepsilon > 0$ small enough, a payoff strictly above α_1, since

$$u_1(\mu, x_2^*) = (1 - \varepsilon)\alpha_1 + \varepsilon u_1(\tilde{x}_1, x_2^*) > 1$$

and

$$u_1(\mu, x_2) = (1 - \varepsilon)\underbrace{u_1(x_1^*, x_2)}_{\underset{\alpha_1}{\vee}} + \varepsilon u_1(\tilde{x}_1, x_2)$$

small

(b)   Take

2	0
0	2
1	1

$\alpha_1 = 1 = v_m < \alpha_2 = 2$

b)   *Finite n-person games*

12)   <u>A game where the mixed equilibrium is not convincing</u>

Consider the 2 x 2 game

0		2	
	1		0
2		1	
	0		1

Compute its (unique) mixed Nash equilibrium (μ_1^e, μ_2^e) and its (unique) prudent mixed strategies (μ_1^P, μ_2^P). Assuming that each player is hesitating between μ_i^e and μ_i^P, draw the corresponding 2 x 2 game and analyze it. Use this to give some strategical advice to each player.

Solution:

$$\mu_1^e = \frac{1}{2}T + \frac{1}{2}B \quad \mu_2^e = \frac{1}{3}L + \frac{2}{3}R$$

$$\mu_1^P = \frac{1}{3}T + \frac{2}{3}B \quad \mu_2^P = \frac{1}{2}L + \frac{1}{2}R$$

	μ_2^e	μ_2^P
μ_1^e	4/3 1/2	5/4 1/2
μ_1^P	4/3 5/9	4/3 1/2

$<=>$

	μ_2^e	μ_2^P
μ_1^e	16 9	15 9
μ_1^P	16 10	16 9

(after rescaling)

μ_1^P is dominating for 1

μ_2^e is dominating for 2

the corresponding outcome is the unanimous optimum. Yet in the original game, Player 2 reacts optimally to μ_1^P by playing R all the time, hence achieving the payoff (2/3) > (5/9) while keeping Player 1's payoff constant at (4/3). (Of course

Player 1 now has an incentive to play Top all the time.)
Good, although morally objectionable, advice might be
Tell Player 2 to play μ_2^e and pretend that you gave
the same advice to 1.

Tell Player 1 to play μ_1^p and pretend Player 2 will
not listen to any advice and is highly unpredictable.

13) Games with infinitely many equilibria

Give an example of a 2 x 2 game where i) no player has
two equivalent pure strategies, ii) there is an infinite
number of Nash equilibria.

Solution:

		L	R	
T	1		2	
		0		1
B	1		0	
		1		0

There are two types of Nash equilibria, namely, (T, R)
and (μ_1, L) for all μ_1 such that $\mu_1(T) \leq (1/2)$.

14) Formula for completely mixed equilibrium

To $G = (X_1, X_2, u_1, u_2)$ associate two $|X_1| \times |X_2|$
matrices $U_i (i = 1, 2)$:

$$U_i = [u_i(x_1, x_2)] \quad x_1 \ \varepsilon \ X_1$$

$$x_2 \ \varepsilon \ X_2$$

Suppose both U_1 and U_2 can be inverted, so that $|X_1| = |X_2|$ and $\det(U_i) \neq 0$ and denote by $\tilde{U}_i(x_1, x_2)$ the cofactor of entry (x_1, x_2) in matrix U_i. Prove that if G has a completely mixed Nash equilibrium, the corresponding payoff to Player i is

$$\frac{\det (U_i)}{\displaystyle\sum_{\substack{x_1 \varepsilon X_1 \\ x_2 \varepsilon X_2}} \tilde{U}_i (x_1, x_2)}$$

Solution:

The payoff to player i can be written as

$$\mu_1 U_i \mu_2 \quad \text{where} \quad \mu_1 \text{ is a row vector}$$

$$\mu_2 \text{ is a column vector}$$

Hence the system (5) (Lemma 4) is written as

$$U_1 \mu_2 = v_1 \mathbb{1}$$

where $\mathbb{1} = (1, \ldots, 1)$ and v_1 is the equilibrium payoff to Player 1.

$$\mu_1 U_2 = v_2{}^t\mathbb{1}$$

where

$$^t\mathbf{1} = \begin{bmatrix} 1 \\ 1 \\ \cdot \\ \cdot \\ \cdot \\ 1 \end{bmatrix}$$

and v_2 is the equilibrium payoff to Player 2.

Since U_i is invertible this gives

$$^t\mu_1 = v_2^t \, \mathsf{U}_2^{-1} \, \mathbf{1} \qquad \mu_2 = v_1 \, \mathsf{U}_1^{-1} \, \mathbf{1}$$

Furthermore, $\mu_1 \, \mathbf{1} = {}^t\mu_2 \mathbf{1} = 1$. Thus

$$v_i = \frac{1}{{}^t\mathbf{1} \, {}^t\mathsf{U}_i^{-1}}$$

It is known that the comatrix $\tilde{\mathsf{U}}$ (whose entries are the cofactors of matrix U) is such that

$$\mathsf{U}^{-1} = \frac{1}{\det(\mathsf{U})} \; {}^t\tilde{\mathsf{U}}$$

Hence

$$^t\mathbf{1} \, {}^t\mathsf{U}_i^{-1} = \frac{1}{\det(\mathsf{U})} \, {}^t\mathbf{1}\tilde{\mathsf{U}}_i = \frac{\sum\limits_{x_1, \, x_2} \tilde{\mathsf{U}}_{i(x_1, \, x_2)}}{\det \mathsf{U}}$$

15) <u>Where Player 1 gains by losing utility</u> (Moulin [1976])

In the first version of the crossing game (Example 6) show that Player 1's payoff from a mixed Nash equilibrium is <u>increased</u> (and Player 2's payoff is unchanged) if Player 1

is penalized (by a small enough amount) for using strategy go.
In other words, prove the following.

$$
G = \begin{array}{|c|c|}
\hline
1 \qquad\quad & 1 - \varepsilon \qquad\quad \\
\qquad\quad 1 & \qquad\quad 2 \\
\hline
2 \qquad\quad & 0 \qquad\quad \\
\qquad\quad 1 - \varepsilon & \qquad\quad 0 \\
\hline
\end{array}
\qquad
G' = \begin{array}{|c|c|}
\hline
1 \qquad\quad & 1 - \varepsilon \qquad\quad \\
\qquad\quad 1 & \qquad\quad 2 \\
\hline
2 - \alpha \qquad\quad & -\alpha \qquad\quad \\
\qquad\quad 1 - \varepsilon & \qquad\quad 0 \\
\hline
\end{array}
$$

For $\varepsilon > 0$, $\alpha > 0$ both small enough, the mixed Nash-equilibrium payoffs in G, G' are such that

$$v_1 < v'_1 \; ; \quad v_2 = v'_2$$

Solution:

For ε, α small enough, both matrices in G' are invertible and the game has a completely mixed equilibrium (in fact $0 < \varepsilon$, $\alpha < 1$ is enough to guarantee this). By the formulas at the end of section 3 we have

$$v^1_1 (\alpha) = \frac{2 - 2\varepsilon + \varepsilon\alpha}{2 - \varepsilon}$$

which is <u>increasing</u> in α

$$v^1_2 (\alpha) = \frac{2(1 - \varepsilon)}{2 - \varepsilon}$$

16) A symmetrical three-person game

In a certain symmetrical three-person game, each player has pure strategies A and B. If Players 1, 2, and 3 play the mixed strategies $(p, 1 - p)$, $(q, 1 - q)$ and $(r, 1 - r)$ respectively, the payoff to Player 1 is

$$u_1(p, q, r) = pqr + 3pq + pr + qr - 2q - p$$

Since the game is symmetrical, there is an equilibrium in which all three players play the same mixed strategy. Find this strategy.

Solution:

We look for some p^*, $0 \leq p^* \leq 1$ such that $u_1(p^*, p^*, p^*) \geq u_1(p, p^*, p^*)$ all $0 \leq p \leq 1$, which is equivalent (since u_1 is linear in its first variable) to

$$u_1(p^*, p^*, p^*) \geq u_1(0, p^*, p^*) \quad \text{and} \quad \geq u_1(1, p^*, p^*)$$

These two give

$$p^{*3} + 5p^{*2} - 3p^* \geq p^{*2} - 2p^* \geq 2p^{*2} + 2p^* - 1$$

Drawing the graph of $y = x^3 + 4x^2 - x = x(x^2 + 4x - 1)$ and $y = x^3 + 3x^2 - 5x + 1 = (x - 1)(x^2 + 4x - 1)$ gives the unique solution $p^* = \sqrt{5} - 2$.

17) <u>Exchangeability of mixed NE outcomes in two-person games</u>
(Parthasarathy-Raghavan [1971])

In a two-person mixed game $G_m = (M_1, M_2, \bar{u}_1, \bar{u}_2)$ the NE outcomes usually are not exchangeable.

$$\mu, \mu' \in NE(G_m) \not\Rightarrow (\mu_1, \mu'_2), (\mu'_1, \mu_2) \in NE(G_m)$$

This should be clear from the crossing game (Example 6).
a) Prove that the NE outcomes of G_m are exchangeable iff $NE(G_m)$ is a convex subset of $R^{X_1} \times R^{X_2}$ (where X_i are, of course, finite). b) If $NE(G_m)$ is indeed a convex (hence rectangular) subset of $M_1 \times M_2$, prove the existence of at least one mixed NE outcome μ^* that Pareto dominates or has the same payoff vector as all other mixed NE outcomes.

$$\forall \mu \in NE(G_m) \quad u_i(\mu) \leq u_i(\mu^*) \qquad i = 1, 2$$

Solution:

(a) Suppose the NE outcomes of G_m are exchangeable and pick $(\mu_1, \mu_2) (\mu'_1, \mu'_2)$ both in $NE(G_m)$. Since (μ_1, μ_2) and (μ'_1, μ_2) are both in $NE(G_m)$, it follows that $u_1(\mu_1, \mu_2) = u_1(\mu'_1, \mu_2) = u_1(\lambda\mu_1 + (1-\lambda)\mu'_1, \mu_2)$ for all $\lambda \in [0, 1]$. Therefore $\mu^\lambda_1 = \lambda\mu_1 + (1-\lambda)\mu'_1$ is a best reply to μ_2. Moreover for any $\tilde{\mu}_2 \in M_2$ we have

$$u_2(\mu^\lambda_1, \mu_2) = \lambda u_2(\mu_1, \mu_2) + (1 - \lambda) u_2(\mu'_1, \mu_2) \geq$$

$$\geq \lambda u_2(\mu_1, \tilde{\mu}_2) + (1 - \lambda) u_2(\mu'_1, \tilde{\mu}_2) = u_2(\mu^\lambda_1, \tilde{\mu}_2)$$

Hence μ_2 is a best reply to μ^{λ}_1. We have proved $(\mu^{\lambda}_1, \mu_2) \in NE(G_m)$. Similarly $(\mu^{\lambda}_1, \mu'_2) \in NE(G_m)$.

Repeat now the argument with μ^{λ}_1 in place of μ_1 and exchanging the roles of Player 1 and 2.

$$\left. \begin{array}{l} (\mu^{\lambda}_1, \mu_2) \in NE(G_m) \\[2ex] (\mu^{\lambda}_1, \mu'_2) \in NE(G_m) \end{array} \right\} \Rightarrow (\mu^{\lambda}_1, \mu^{\lambda}_2) = (\mu^{\lambda}_1, \lambda\mu_2 + (1 - \lambda)\mu'_2) \in NE(G_m)$$

To prove the converse statement, suppose $NE(G_m)$ is convex and pick (μ_1, μ_2) (μ'_1, μ'_2) two mixed equilibria. For any $\lambda \in [0, 1]$, we have $(\mu^{\lambda}_1, \mu^{\lambda}_2) \in NE(G_m)$ hence:

$$u_1(\mu^{\lambda}_1, \mu^{\lambda}_2) \geq u_1(\mu_1, \mu^{\lambda}_2), \ u_1(\mu^{\lambda}_1, \mu^{\lambda}_2) \geq u_1(\mu'_1, \mu^{\lambda}_2)$$

For λ different of zero and 1 this yields

$$u_1(\mu^{\lambda}_1, \mu^{\lambda}_2) = u_1(\mu_1, \mu^{\lambda}_2) = u_1(\mu'_1, \mu^{\lambda}_2) = \sup_{\mu_1} u_1(\tilde{\mu}_1, \mu^{\lambda}_2)$$

By continuity we can take $\lambda = 0$ or 1 in this equation, implying that μ'_1 is a best reply to μ_2 and μ_1 is a best reply to μ'_2. Symmetrical arguments yield μ'_2 is a best reply to μ_1 and μ_2 a best reply to μ'_1.

(b) Let $NE(G_m) = N_1 \times N_2$. Here N_1, N_2 are both convex and compact and:

$$\forall \ \mu_1, \mu'_1 \in N_1 \ \forall \ \mu_2 \in N_2: \ u_1(\mu_1, \mu_2) = u_1(\mu'_1, \mu_2)$$

Thus on $N_1 \times N_2$ u_1 depends on $\mu_2 \, \epsilon \, N_2$ only and u_2 depends on $\mu_1 \, \epsilon \, N_1$ only. Choose μ_1^* maximizing $u_2(\mu_1, \cdot)$ on N_1 and μ_2^* maximizing $u_1(\cdot, \mu_2)$ on N_2. The equilibrium (μ_1^*, μ_2^*) is Pareto dominant.

c) *Infinite games*

18) <u>Picking integers</u>

Players 1 and 2 each pick a positive integer. The player with the lower integer wins \$1 from the other unless he is exactly one lower, in which case he loses \$2. If they both pick the same number, the payoff is zero to both. Find the unique optimal mixed strategy (the value is zero by symmetry). <u>Hint</u>: The carrier of an optimal strategy is finite.

Solution:

The matrix is

	1/16	5/16	4/16	5/16	1/16
1/16	0	-2	1	1	1
5/16	2	0	-2	1	1
4/16	-1	2	0	-2	1
5/16	-1	-1	2	0	-2
1/16	-1	-1	-1	2	0

Optimal strategies are shown on the figure. The uniqueness
of this equilibrium holds because any other equilibrium
strategy must guarantee a nonnegative payoff to Player 1
against the above optimal strategy μ_2^*. Writing this condition
shows that no pure strategy above $x_1 = 5$ can receive a positive
weight.

19) <u>Hiding a number</u>

Two players pick a positive integer. If these two numbers
differ, the payoff is zero. If they coincide, $x_1 = x_2 = p$,
Player 1 receives a_p from Player 2 where $a_1 \leq a_2 \leq \ldots \leq a_p \leq \ldots$
is a nondecreasing sequence of positive numbers. If they do not,
the payoff is zero to both. Thus, Player 2 faces a dilemma.
The larger his number, the less probable it is that Player 1 can
guess it, but the more risky a potential discovery. a) Write
the normal form of the pure and mixed games. Observe that the
payoff might be $+\infty$. b) Suppose first

$$\sum_{p=1}^{+\infty} \frac{1}{a_p} = \frac{1}{\alpha} < +\infty$$

Prove that there is a unique, symmetric, completely mixed
saddle point and the value is α. c) Suppose next

$$\sum_{p=1}^{+\infty} \frac{1}{a_p} = +\infty$$

Prove that the value is zero, that every strategy of Player 1

is optimal, and yet Player 2 has no optimal strategy. Describe
an ε optimal strategy for Player 2.

Solution:

(a) Here $M_1 = M_2 = \{\mu \in R^N / \sum_{p=1}^{+\infty} \mu(p) = 1$ and $\mu(p) \geq 0$ all $p \in N\}$,
i.e., the set of probability distributions over N.

$$u_1(\mu_1, \mu_2) = \sum_{p=1}^{+\infty} a_p \cdot \mu_1(p) \cdot \mu_2(p)$$

which is possibly infinite.

(b) Define μ^*: $\mu^*(p) = \dfrac{\alpha}{a_p} > 0$, $p = 1, 2, \ldots$
Clearly $\mu^* \in M_i$ and

$$u_1(\mu_1, \mu^*) = u_1(\mu^*, \mu_2) = \alpha \qquad \text{all } \mu_1, \mu_2$$

Hence μ^* is the completely mixed optimal strategy of both
players.

(c) We prove now

$$\inf_{\mu_2} \sup_{\mu_1} \bar{u}_1(\mu_1, \mu_2) = 0 \qquad\qquad (1)$$

As there are no negative payoffs we have $0 \leq \sup_{\mu_1} \inf_{\mu_2} \bar{u}_1$.
This implies that the value exists and is zero. To prove
equation (1) pick any $\varepsilon > 0$ and construct a strategy μ_2
such that

$$\sup_{\mu_1} \bar{u}_1(\mu_1, \mu_2) = \sup_{p \in N} \bar{u}_1(\delta_p, \mu_2) = \sup_p a_p \cdot \mu_2(p) \leq \varepsilon$$

118

This amounts to the inequalities

$$0 \le \mu_2(p) \le \frac{\varepsilon}{a_p} \qquad \text{all } p \in N$$

$$\sum_p \mu_2(p) = 1$$

which is clearly possible since $\sum_p (\varepsilon/a_p) = +\infty$. Proving
that Player 2 has no optimal strategy, we have for all μ_2

$$\sup_{\mu_1} \bar{u}_1(\mu_1, \mu_2) = \sup a_p \cdot \mu_2(p) > 0$$

Since a_p are all nonzero and some $\mu_2(p)$ must be positive.

20) <u>Catch me</u>

a) Player 2 picks a number x_2 in $[0, 1]$. Player 1 tries
to catch him by picking an x_1 in $[0, 1]$ within a distance 1/4:

$$X_1 = X_2 = [0, 1] \quad u_1(x_1, x_2) = 1 \quad \text{if } |x_1 - x_2| \le \frac{1}{4}$$

$$= 0 \quad \text{if } |x_1 - x_2| > \frac{1}{4}$$

In the pure-strategy game, we have $\alpha_1 = 0 < 1 = \alpha_2$. Prove
that the mixed extension of the game has value 1/2 and compute
a pair of optimal strategies.

<u>Hint</u>: Seek optimal mixed strategies where each player uses
only two pure strategies.

b) Consider now the variant of the game of catch me where

$$u_1(x_1, x_2) = 1 \qquad \text{if } |x_1 - x_2| < \frac{1}{4}$$

$$= 0 \qquad \text{if } |x_1 - x_2| \ge \frac{1}{4}$$

Although the change in the payoff function is small, the value decreases to 1/3. Explain why by giving a pair of optimal mixed strategies.

Solution:

(a) A pair of optimal mixed strategies is

$$\mu_1 = \frac{1}{2}\delta_{\frac{1}{4}} + \frac{1}{2}\delta_{\frac{3}{4}}, \quad \mu_2 = \frac{1}{2}\delta_0 + \frac{1}{2}\delta_1$$

Player 1's optimal strategy is unique, but not Player 2's (for instance $\mu_2^1 = \frac{1}{2}\delta_{.1} + \frac{1}{2}\delta_{.9}$ is optimal as well). The value is 1/2.

(b) Now Player 2 has a unique optimal strategy $\mu_2 = \frac{1}{3}\delta_0 + \frac{1}{3}\delta_{\frac{1}{2}} + \frac{1}{3}\delta_1$ guaranteeing that he is caught with probability at most $\frac{1}{3}$. On the other hand Player 2 has many optimal strategies, e.g., $\mu_1 = \frac{1}{3}\delta_{\frac{1}{8}} + \frac{1}{3}\delta_{\frac{1}{2}} + \frac{1}{3}\delta_{\frac{7}{8}}$. The value is 1/3.

21) First-price, auction dollar game

In a variant of the auction dollar game (Example 2, Chapter 2) the highest bidder gets the prize and every player pays his **own** bid

$$u_i(x) = 1 - x_i \qquad \text{if } x_i > x(i)$$
$$= -x_i \qquad \text{if } x_i < x(i)$$

where $x(i) = \sup_{j \neq i} x_j$.

Prove that every strategy $x_i > 1$ is dominated. Then compute the symmetrical mixed-Nash equilibrium with a positive density on $[0, 1]$.

Hint: See Example 7.

Solution:

Strategy 1 dominates every $x_i > 1$, as one checks easily. We seek now a probability distribution γ with carrier $[0, 1]$ such that $\bar{u}_i(\delta_{x_i}, \mu_{-i})$ is independent of $x_i \in [0, 1]$ whenever $\mu_j = \gamma$ for all $j \neq i$. Denote by F the cumulative distribution of γ. Then compute

$$u_i(\delta_{x_i}, \mu_{-i}) = (1 - x_i) \text{ Prob } (x(i) < x_i) - x_i \text{ Prob } (x_i < x(i))$$

$$= (1 - x_i) F^{n-1}(x_i) - x_i (1 - F^{n-1}(x_i))$$

$$= F^{n-1}(x_i) - x_i$$

Thus the property $(\partial \bar{u}_i / \partial x_i) = 0$ amounts to $F(x) = 1/x^{n-1}$, $0 \leq x \leq 1$. The corresponding density is $\dfrac{1}{n-1} \cdot x^{-\frac{n-2}{n-1}}$, $0 \leq x \leq 1$. The equilibrium payoff is zero to all players.

22) A location game

Two shopowners must decide the location of their respective shops along an interval $[0, 1]$. Player 1 sells cheap sports equipment, whereas Player 2 deals in elegant sports gear. As side-by-side comparison is, on average, favorable to the merchant of cheap equipment, the players face an inelastic demand; Player 1 wants to locate as close as possible to Player 2, whereas Player 2 tries to move as far as possible from Player 1. We assume that the profit functions take the following form.

$$X_1 = X_2 = [0, 1]$$

$$u_1(x_1, x_2) = 1 - |x_1 - x_2|$$

$$u_2(x_1, x_2) = |x_1 - x_2| \qquad \text{if } |x_1 - x_2| \leq \frac{2}{3}$$

$$= \frac{2}{3} \qquad \text{if } |x_1 - x_2| \geq \frac{2}{3}$$

Notice the negative externalities imposed by Player 1 on Player 2 vanish when their distance is at least 2/3.

The pure-strategy game has no Nash equilibrium. Show that Glicksberg's theorem implies the existence of a mixed Nash equilibrium. Check that the following pair is a Nash equilibrium.

$$\mu^*_1 = \frac{1}{3} \delta_0 + \frac{1}{6} \delta_{\frac{1}{3}} + \frac{1}{6} \delta_{\frac{2}{3}} + \frac{1}{3} \delta_1$$

$$\mu^*_2 = \frac{1}{2} \delta_0 + \frac{1}{2} \delta_1$$

Neither of these two mixed strategies are completely mixed. However (μ^*_1, μ^*_2) share the typical property of completely mixed equilibria, namely,

$$\bar{u}_1(\mu_1, \mu^*_2) = \bar{u}_1(\mu^*_1, \mu^*_2) \qquad \text{all } \mu_1 \in M_1$$

$$\bar{u}_2(\mu^*_1, \mu_2) = \bar{u}_2(\mu^*_1, \mu^*_2) \qquad \text{all } \mu_2 \in M_2$$

Solution:

To check that $\bar{u}_1(\mu_1, \mu_2^*)$ is independent of μ_1, it is enough to check that $\bar{u}_1(\delta_{x_1}, \mu_2^*)$ is independent of the <u>pure</u> strategy $x_1 \in [0, 1]$.

$$\bar{u}_1(\delta_{x_1}, \mu_2^*) = \frac{1}{2}(1 - x_1) + \frac{1}{2}(1 - (1 - x_1)) = \frac{1}{2}$$

Similarily

$$\bar{u}_2(\mu_1^*, \delta_{x_2}) = \frac{1}{3}x_2 + \frac{1}{6}(\frac{1}{3} - x_2) + \frac{1}{6}(\frac{2}{3} - x_2) + \frac{1}{3} \cdot \frac{2}{3} \qquad \text{if } 0 \le x_2 \le \frac{1}{3}$$

$$= \frac{1}{3}x_2 + \frac{1}{6}(x_2 - \frac{1}{3}) + \frac{1}{6}(\frac{2}{3} - x_2) + \frac{1}{3}(1 - x_2) \qquad \text{if } \frac{1}{3} \le x_2 \le \frac{2}{3}$$

$$= \frac{1}{3} \cdot \frac{2}{3} + \frac{1}{6}(x_2 - \frac{1}{3}) + \frac{1}{6}(x_2 - \frac{2}{3}) + \frac{1}{3}(1 - x_2) \qquad \text{if } \frac{2}{3} \le x_2 \le 1$$

Therefore $\bar{u}_2(\mu_1^*, \mu_2) = (7/18)$ all $\mu_2 \in M_2$. Remark that μ_2^* is an optimal strategy for Player 1 in the zero-sum game $(M_1, M_2, \bar{u}_1)$ since $\bar{u}_1(\mu_1, \mu_2) = \bar{u}_1(\mu_2, \mu_1)$ for all μ_1, μ_2. Thus at the Nash equilibrium (μ_1^*, μ_2^*), Player 1 gets only his secure utility level.

CHAPTER 8   CORRELATED EQUILIBRIUM

1) The crossing game

In the crossing game of Example 1 find the best symmetrical correlated equilibrium.

Solution:

A symmetrical lottery takes the form

$$L = \begin{array}{|c|c|} \hline p & q \\ \hline q & 0 \\ \hline \end{array} \quad p, q \geq 0, \quad p + 2q = 1$$

It is a correlated equilibrium iff

124

$$p \cdot 1 + q \cdot (.5) \geq p(1.3) + q(0)$$

<table>
<tr><td>expected payoff
if 1 is told to
stop and stops</td><td>expected payoff
if 1 is told to
stop but does not</td></tr>
</table>

We want to maximize $p + q(.5) + q(1.3) = p(.1) + (.9)$ under this constraint. This amounts to

$$\max (.1)p + (.9) = \frac{1}{22} + \frac{9}{10} \approx .95$$

$$\text{s.t.} \ (.55)p \leq .25$$

and the best symmetrical correlated equilibrium has $p = (1/2.2) \approx .46$, $q \approx .27$.

2) 2 x ... x 2 games

In a 2 x ... x 2 game (each player has exactly two strategies) prove that Definition 1 and 2 coincide.

Solution:

This follows from the mere comparison of Definition 1 and 2.

3) Musical chairs

In the musical-chair game (Example 2) prove that the sets $NE(G)$, $NE(G_m)$, $CE(G)$ and $WCE(G)$ are all different. Prove that in a Pareto-optimal CE payoff, a player receives no more than 5/3 and no less than 4/3 (and both bounds are reached).

However, in a Pareto-optimal WCE a player's payoff can be anywhere between 1 and 2.

Solution:

In the musical chairs game we know already NE(G) = ϕ and that NE(G_m) is a Pareto-dominated singleton. We compute now the Pareto-optimal correlated equilibria.

A lottery L is Pareto-optimal iff it gives probability zero to the diagonal entries of the game. By the convexity of CE and symmetry of the game, the Pareto-optimal CE payoffs cover an interval $[(\lambda, 3 - \lambda), (3 - \lambda, \lambda)]$. To compute λ we find the CE most favorable to Player 1. By symmetry of the game and convexity of CE, there exists one such CE of the form

$$L = \begin{array}{|c|c|c|}
\hline
0 & p & q \\
\hline
q & 0 & p \\
\hline
p & q & 0 \\
\hline
\end{array} \qquad p + q = \frac{1}{3} \qquad q \geq \frac{1}{2}$$

The equilibrium inequalities reduce to

Player 1: $p + 2q \geq 2p$

Player 2: $q + 2p \geq 2q$

So the program

$$\max\ 3p + 6q$$

$$\text{s.t.}\ \frac{p}{2} \le q \le 2p$$

has optimal solution $p^* = 1/9$, $q^* = 2/9$ and value $5/3$.

On the other hand <u>any</u> lottery such as L is a weak correlated equilibrium since the equilibrium inequalities are just:

Player 1: $3p + 6q \ge 1(\frac{1}{3}) + 2(\frac{1}{3})$

Player 2: $6p + 3q \ge 1(\frac{1}{3}) + 2(\frac{1}{3})$

So the best CE for Player 1 has payoffs $(5/3)$, $(4/3)$ as was to be shown.

CHAPTER 9   COALITIONAL STABILITY: THE CORE

a)   *Imputations and a core*

1)   <u>A three-player, zero-sum game</u>

Three players 1, 2, 3 rotate taking one or two sticks from
a pile originally containing n ≥ 1 sticks. Player 1 goes first
followed by Player 2, then by Player 3, then by Player 1 again
(if the pile is not exhausted) and so on. Say that i is the
player who takes the last stick; then, the payoff is

-3 to Player i

+2 to Player i + 1

+1 to Player i + 2

(where we set 3 + 1 = 1, 3 + 2 = 2, 2 + 2 = 1).

a) Compute the perfect equilibrium of this game for all n.
b) Prove that coalition {12} can force 3 to lose except for
n = 1, 2, 6, coalition {23} can force 1 to lose except for
n = 2, 3, 4, 7, 8, coalition {13} can force 2 to lose except
for n = 1, 4, 5, 9.   c) Deduce that for n ≥ 10 the α core
of the game is empty.   Compute it for 3 ≤ n ≤ 9.

Solution:

    (a)   For n = 4k + 1, Player 1 loses: payoffs are (-3, 2, 1)

            For n = 4k + 2, Player 2 loses: payoffs are (1, -3, 2)

            For n = 4k + 3 or = 4k, Player 3 loses: payoffs
are (2, 1, -3).

    Check first the claim for n = 1, 2, 3, 4.   Then proceed
by an induction argument:

    · If n = 4k + 1 no matter what Player 1 plays we go to
a position where he (now the third player) loses

    · If n = 4k + 2 Player 1 has a choice of forcing 2 to lose
(by removing 1 stick) or make himself the loser (by removing
two sticks).

    · If n = 4k + 3 Player 1 can force 3 to lose (by removing
one stick) or force 2 to lose (removing two sticks).   He
prefers the former.

    · If n = 4k + 4 Player 1 makes himself the loser if he
removes one stick.   He then prefers to remove two sticks and
make Player 3 the loser.

129

(b) If a coalition forms against Player i and he must play in a pile of $4k + 1$, $k \geq 0$, he loses, since the other two players can force his next pile to be $4(k - 1) + 1$ and so on. Next if Player i faces a pile of $4k + 2$, $k \geq 1$, he loses since after his play (going to $4k + 1$ or $4k$) the coalition can leave him with $4(k - 1) + 1$ in his next turn.

Next, if Player i faces a pile of $4k + 3$ or $4k + 4$, and $k \geq 2$, he loses, since no matter how he plays the coalition can bring him back to $4(k - 1) + 2$ or $4(k - 1) + 1$ or $4k + 1$ at his next turn.

It remains to fill the blanks left by this argument for small values of n and the three two-players coalition. This will lead the careful reader to the announced statement.

(c) For $3 \leq n \leq 9$, at most two coalitions can make the third player lose. Without loss of generality suppose they are 12 and 23.

Outcomes with payoff $(1, -3, 2)$ are blocked by the coalition {12}

Outcomes with payoff $(2, 1, -3)$ are blocked by the coalition {23}

Outcomes with payoff $(-3, 2, 1)$ are not blocked

Thus the α core is when Player 1 loses. Plugging this into the structure of winning coalitions discovered in b) yields that the α core outcomes are the <u>same</u> as the sophisticated outcomes for $3 \leq n \leq 9$. For $n \geq 10$, the α core is empty.

2) <u>Symmetrical games have a nonempty α core</u>

Let G be a symmetrical game. For any permutation σ of $\{1, \ldots, n\}$ and any strategy n-tuple $x \in X_N$, denote x^σ the following strategy n-tuple:

$$(x^\sigma)_i = x_{\sigma(i)}$$

for all $i = 1, \ldots, n$. Then we have

$$u_{\sigma(i)}(x^\sigma) = u_i(x)$$

for all $i = 1, \ldots, n$, and all $x \in X_N$.

Suppose G has compact strategy sets and continuous payoff functions. Prove that the α core of G is nonempty.

Solution:

By induction on the number n of players. Suppose the claim holds until $n - 1$. Let G be an n players game. Define $G(-n)$ to be the game with players $1, \ldots, n - 1$ and payoffs

$$\tilde{u}_i(x_{-n}) = \inf_{x_n} u_i(x_{-n}, x_n)$$

By induction assumption, the α core of $G(-n)$ is nonempty since $G(-n)$ is symmetrical. Also, we can pick some $x^*_{-n} \in C_\alpha(G(-n))$

such that $\tilde{u}_i(x^*_{-n}) \le \tilde{u}_{n-1}(x^*_{-n})$ for all $i = 1, \ldots, n - 2$.
Then consider the outcome $x^* = (x^*_{-n}, x^*_n)$ of G where
$x^*_n = x^*_{n-1}$. We claim that no coalition of size at most
$(n - 1)$ has an objection against x^*. Namely, if a coalition
$S \subset \{1, \ldots, n - 1\}$ has a strategy x_S such that

$$\inf_{x_{N \backslash S}} u_i(x_S, x_{N \backslash S}) > u_i(x^*) \qquad \text{all } i \in S$$

then

$$\inf_y \tilde{u}_i(x_S, y_{N \backslash S - n}) > u_i(x^*) \ge \tilde{u}_i(x^*_{-n}) \qquad \text{all } i \in S$$

This would contradict the assumption that $x^*_{-n} \in C_\alpha(G(-n))$.
Similarly, coalition S contained in $\{1, \ldots, (n - 2), n\}$
cannot have an objection against x^* (since players n and
$(n - 1)$ use the same strategy in x^* and G is symmetrical).
Finally suppose that a coalition S containing n, $(n - 1)$
but not containing Player 1, say, has an objection against x^*.
Let S' be the same coalition except that Player 1 replaces
Player n. By symmetry of G, S' has the same strategic
opportunities as S; by construction of x^*, $\tilde{u}_1(x^*_{-n}) \le$
$\tilde{u}_{n-1}(x^*_{-n}) \le u_{n-1}(x^*) = u_n(x^*)$, so that S' would raise an
objection against x^*_{-n} in G(-n). We have proved that no
proper coalition will object in G against x^*. It remains
to choose a Pareto-optimal outcome x of G, such that
$u_i(x) \ge u_i(x^*)$ for all $i = 1, \ldots, n$. This outcome is in
the α core of G.

b) *Games in characteristic form*

3) A symmetrical game

A symmetrical n-person game has $v(S) = -1$ for all coalitions
S. a) Prove the core is nonempty. b) Define it by $n + 1$
inequalities instead of $2^n - 1$.

Hint: Start with $n = 3$, 4 and generalize.

c) Describe its most unfair elements. (By "unfair" we mean
that the difference between the richest and poorest players
is as large as possible).

Solution:

A vector x is in the core if

$$\sum_{i \in S} x_i \geq -1 \qquad \text{all } S \subset N \text{ and } \sum_{i \in N} x_i = -1$$

which is equivalent to

$$\sum_{i \in N} x_i = -1 \text{ and } x_i \leq 0 \quad \text{for all } i \in N$$

Hence, the most unfair elements are the distributions such as
$(-1, 0, \ldots, 0)$.

4) Pollute the lake

There are n factories around a lake. It costs B (per day)
for a factory to treat its wastes before discharging then
into the lake. It costs kA (per day) for a factory to purify
its water supply, if k is the number of factories that <u>do not</u>

treat their wastes. Assume $A \leq B \leq nA$. a) Determine the characteristic function. b) When does this game have a core? Does it ever have a 1-point core? If so, when?

Solution:

(a) Consider first the grand coalition N. If a player switches from "do not purify" to "purify" his contribution to the joint utility switches from $-nA$ to $-B$. By assumption $B \leq nA$, so it pays to purify at the level of coalition N: $v(n) = -nB$.

Similarly, it pays to purify at the level of a coalition of size t if $B \leq tA$. Hence, the characteristic function

$$v(t) = -t\{(n - t)A + \min(B, tA)\}$$

Notice that coalitions assume the worst from players in $N\backslash T$.

The core is nonempty since

$$\frac{v(n)}{n} = -B \geq -(n - t)A - \min\{B, tA\} = \frac{v(t)}{t}$$

It has a 1-point core (namely $(1/n, \ldots, 1/n)$) iff one of the inequalities $(v(n)/n) \geq (v(t)/t)$ is an equality for some $t < n$, which is possible iff $B = nA$.

5) <u>An interesting core</u>

Assume that n is <u>even</u> and consider the n-person game with $v(N) = n/2$,

$$v(1, 2) = v(3, 4) = \ldots = v(n - 3, n - 2) = v(n - 1, n) = 1$$

$$v(2, 3) = v(4, 5) = \ldots = v(n - 2, n - 1) = v(n, 1) = 1$$

and $v(S)$ is the minimum determined by superadditivity for all other $S \subset N$. Show that the core is the line segment between the "odd" imputation $(1, 0, 1, 0, \ldots, 1, 0)$ and the "even" imputation $(0, 1, 0, 1, \ldots, 0, 1)$.

Solution:

First check that the odd and even imputations are in the core. It is enough to check this for the two person coalitions with $v(ij) = 1$, since the characteristic function is derived from those by superadditivity.

Now suppose x is an element in the core. Adding up the inequalities

$$x_1 + x_2 \geq v(1, 2) = 1$$
$$x_3 + x_4 \geq v(3, 4) = 1$$
$$\vdots$$
$$x_{n-1} + x_n \geq v(n - 1, n) = 1$$

yields an equality since $v(N) = (n/2)$. Hence all these are equalities. By the same token, one proves $x_2 + x_3 = v(2, 3)$, $x_4 + x_5 = v(4, 5), \ldots, x_n + x_1 = v(n, 1)$.

Hence the system

$$x_1 + x_2 \qquad\qquad\qquad\qquad = 1$$
$$x_2 + x_3 \qquad\qquad\qquad = 1$$
$$x_3 + x_4 \qquad\qquad = 1$$
$$\cdots$$
$$x_{n-1} + x_n = 1$$
$$x_1 \qquad\qquad\qquad x_n = 1$$

Solutions of this system form the straight line between the odd and the even imputation.

6) Inessential games

Say that the characteristic function v is inessential if its imputation set is a singleton:

$$v(N) = \sum_{i \in N} v(i)$$

Show that a game is {superadditive and inessential} iff it is additive:

$$\text{for all coalition } S: \quad v(S) = \sum_{i \in S} v(i)$$

Solution:

Let v be superadditive and inessential. Fix a coalition S. By superadditivity

$$v(S) \geq \sum_{i \in S} v(i)$$

$$v(N \setminus S) \geq \sum_{i \in N \setminus S} v(i)$$

$$v(N) \geq v(S) + v(N \setminus S)$$

Summing these unequalities gives an equality (v is inessential) so that we have equality everywhere, and v is additive. The converse statement is obvious.

7) <u>Convex games</u> (Shapley [1971])

Say that (N, v) is a convex game if

$$v(S) + v(T) \leq v(S \cap T) + v(S \cup T)$$

for all S, T $\subset$ N.

a) Prove in this case

$$v(N) - v(S^c) = \sup_{T \subseteq S^c} \{v(S \cup T) - v(T)\}$$

for all S $\subset$ N.

b) For any ordering 1, 2, ..., n of N, set

$$x_i = v(1, 2, \ldots, i) - v(1, 2, \ldots, i - 1), \quad i = 2, \ldots, n$$

$$x_1 - v(1)$$

Prove that x is in the core.

c) In fact, C(v) is the convex hull of the imputations x_σ obtained as in b) when the ordering σ of N varies. Prove this for n = 3.

Solution:

(a) Apply convexity to (S $\cup$ T) and S^c where S, T are any two disjoint coalitions:

$$v(S \cup T) + v(S^c) \leq v(T) + v(N)$$

$$T = (S \cup T) \cap S^c \qquad N = (S \cup T) \cup S^c$$

This proves

$$v(N) - v(S^c) \geq \sup_{T \subseteq S^c} v(S \cup T) - v(T)$$

For the converse inequality take $T = S^c$.

(b) Fix a coalition S and order its elements as

$$S = \{i_1, i_2, \ldots, i_{|S|}\}$$

Then compute

$$x_{i_1} = v(1, \ldots, i_1) - v(1, \ldots, i_1 - 1) \geq v(i_1)$$

(by superadditivity, a consequence of convexity).

$$x_{i_1} + x_{i_2} \geq v(i_1) + v(1, \ldots, i_2) - v(1, \ldots, i_2 - 1) \geq v(i_1, i_2)$$

(by convexity).

$$x_{i_1} + x_{i_2} + x_{i_3} \geq v(i_1, i_2) + v(1, \ldots, i_3) - v(1, \ldots, i_3 - 1) \geq$$

$$v(i_1, i_2, i_3)$$

(by convexity again). And so on, until finally,

$$x_{i_1} + \ldots + x_{i_{|S|}} \geq v(i_1, \ldots, i_{|S|})$$

which was to be proved.

(c)　A superadditive, noninessential, three-players game can be normalized as in Example 5

$$v(i) = 0 \quad i = 1, 2, 3 \quad v(123) = 1$$

$$v(12) = a_3, \quad v(23) = a_1, \quad v(31) = a_2, \quad 0 \le a_i \le 1$$

Convexity of v means

$$a_1 + a_2 \le 1, \quad a_2 + a_3 \le 1, \quad a_3 + a_1 \le 1$$

The six vectors isolated in question (b) are

$$\alpha^1 = (0, a_3, 1 - a_3), \quad \beta^1 = (0, 1 - a_2, a_3), \quad \alpha = (a_3, 0, 1 - a_3),$$

$$\beta^2 = (1 - a_1, 0, a_1), \quad \alpha^3 = (a_2, 1 - a_2, 0), \quad \beta^3 = (1 - a_1, a_1, 0)$$

We know they are all in the core.　We must show their convex hull is the core.

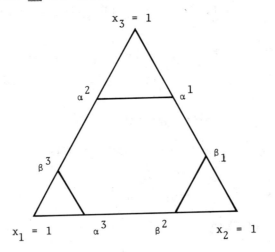

Indeed $\alpha^1\alpha^2$ is the line $x_1 + x_2 = a_3$, $\beta^2\alpha^3$ is the line $x_2 + x_3 = a_1$, and $\beta^3\beta^1$ is the line $x_1 + x_3 = a_2$. Thus, outside the hexagon spanned by α^i, β^i, there are no points in the core. However, the core is convex, so it contains our hexagon.

8) <u>Simple games</u>

We say that game (N, v) is <u>simple</u> if

$$\text{for all } S \subseteq N, \ v(S) = 0 \text{ or } 1$$

In that case we denote by W the set of winning coalitions, namely:

$$S \in W \iff v(S) = 1$$

a) Prove that the simple game (N, v) is superadditive iff W is monotonic and proper:

$$\text{monotonic: } S \in W, \ S \subseteq T \implies T \in W \quad \text{for all } S, T$$

$$\text{proper: } S \in W \implies S^c \notin W$$

From now on, assume that W is superadditive. b) Say that Player i* $\in$ N is a dictator of W if {i*} is a winning coalition. Assuming that (N, v) is superadditive, prove that W has a (unique) dictator iff it is inessential. c) Suppose that W has no dictator and is nonempty. Say that Player i* is a <u>veto player</u> if $N\setminus\{i*\}$ is <u>not</u> winning. Denote by N_* the possibly empty set of veto players. Prove that the core $C(v)$ is nonempty iff N_* is nonempty. In this case $x \in C(v)$ iff

$$\sum_{i \in N_*} x_i = 1, \; x_i = 0 \qquad \text{for all } i \notin N_*$$

$$x_i \geq 0 \qquad \text{for all } i \in N_*$$

d) Let for all i, Player i be endowed with a voting weight $q_i \geq 0$ and denote by $q_o \geq 0$ a quota such that

$$q_o \leq \sum_{i \in N} q_i < 2q_o$$

Then define the weighted majority game W_q by

$$S \in W_q \iff \sum_{i \in S} q_i \geq q_o$$

Prove that W_q is a monotonic and proper simple game. Prove that i* is a dictator iff $q_o \leq q_{i*}$ and i* is a veto player iff

$$\sum_{i \in N} q_i - q_o < q_{i*}$$

Solution:

(a) If v is simple, it is supperadditive iff

{S and T disjoint and $v(S) = v(T) = 1$} is impossible

{S and T disjoint, $v(S) = 1$, $v(T) = 0$} implies

$\{v(S \cup T) = 1\}$

The first property amounts to properness, the second to monotonicity.

(b) Since v is simple, it is inessential iff

$$\sum_{i \in N} v(i) = 1 \iff \text{there is a unique i such that } v(i) = 1$$

(c) By definition, i* is a veto player iff $v(N \backslash i^*) = 0$. Suppose i* is a veto player. Then the imputation $x_{i^*} = 1$, $x_i = 0$ all $i \neq i^*$ is in $C(v)$. Namely for each S, $x(S) \geq v(S)$ is no problem if $i^* \in S$ since $x(S) = 1$. If $i^* \notin S$, then

$$v(S) \leq v(N^* \backslash i) = 0 \Rightarrow v(S) = 0$$

Conversely suppose $x \in C(v)$ and the game has no veto player. So, for all $i \in N$, $v(N \backslash i) = 1$. Thus we must have $1 - x_i = x(N \backslash i) \geq v(N \backslash i) \Rightarrow x_i \leq 0$, for all i., Contradiction.

Suppose next N_* is nonempty and $x \in C(v)$. By the same reasoning we get $x_i \leq 0$ for all $i \notin N_*$. Hence $x_i = 0$ for all $i \in N_*$. Thus, $x(N_*) = 1$. Conversely, let x be such that $x_i = 0$ if $i \notin N_*$, $x_i \geq 0$ if $i \in N_*$, and $x(N_*) = 1$. If S is a coalition, observe that $v(S) = 0$ unless $N_* \subset S$. If $j \in N_*$, $j \notin S$: $v(S) \leq v(N \backslash j) = 0$. Therefore x is in the core of our game.

(d) That W is monotonic is obvious. Properness follows from

$$\text{S and T disjoint, } \sum_{i \in S} q_i \geq q_0, \text{ and } \sum_{i \in T} q_i \geq q_0 \Rightarrow \sum_{i \in N} q_i \geq 2q_0$$

The last two statements are just as simple.

CHAPTER 10   THREATS AND REPETITION

a) *Strong equilibrium and a core*

1) A location game

Two shopowners choose the location of their respective
shops along the [0, 1] interval. They supply complementary
goods (say sports equipment and travel-agent services),
implying a positive externality from either shop to the
other. Moreover Player 1 inclines to be located as close
to zero as possible, whereas Player 2 seeks to be located as
far as possible from zero.

Specifically, they face the following game.

$$X_1 = X_2 = [0, 1]$$

$$u_1(x_1, x_2) = \alpha_1 x_1 - |x_1 - x_2|$$

$$u_2(x_1, x_2) = \alpha_2(x_2 - 1) - |x_1 - x_2|$$

where $\alpha_1 < 0 < \alpha_2$

Suppose that $|\alpha_i| \leq 1$, $i = 1, 2$. a) Compute the best-reply correspondences of both players and the Nash equilibrium. Discuss their stability. b) Find the strong equilibrium outcomes.

Solution:

(a) Best-reply functions are

$$r_i(x_j) = x_j \qquad\qquad \text{all } x_j \in [0, 1]$$

Thus the Nash equilibria are all outcomes (λ, λ), $0 \leq \lambda \leq 1$. None of them is stable. Starting from (λ, μ) with $\mu \neq \lambda$, the simultaneous Cournot tatonnement is $(\mu, \lambda)(\lambda, \mu)(\mu, \lambda)$, and so on. Yet, a Nash equilibrium (λ, λ) is locally stable in a weak sense. If Player 1 deviates slightly to λ^1, the sequential $\{2, 1\}$ tatonnement (see Exercise 4, Chapter 6) converges to (λ^1, λ^1) at once. Thus, a slight deviation implies a slight change of the equilibrium.

(b) Compute first the Pareto-optimal outcomes. If $x_1 \neq x_2$, (x_1, x_2) is Pareto inferior to (λ, λ) where

$\lambda = (1/2)(x_1 + x_2)$:

$$\alpha_1 x_1 - |x_1 - x_2| < \frac{1}{2}\alpha_1(x_1 + x_2) \qquad \text{since } |\alpha_1| < 1$$

$$\alpha_2(x_2 - 1) - |x_1 - x_2| < \alpha_2(\frac{1}{2}(x_1 + x_2) - 1) \text{ since } |\alpha_2| < 1$$

However, (λ, λ) is Pareto-optimal since $\alpha_1 \cdot \alpha_2 < 0$. Thus the Nash equilibria are the strong equilibria as well.

2) A three-player prisoners' dilemma

Each player can be aggressive (A) or cooperative (C) and the game is symmetrical. The various payoffs are listed below.

> (C, C, C) payoff vector (2, 2, 2)
>
> (A, C, C) payoff vector (3, 1, 1)
>
> (A, A, C) payoff vector (2, 2, 0)
>
> (A, A, A) payoff vector (1, 1, 1)

For instance, the three players are three firms competing by setting a regular price (C) or using a dumping policy (A). The maximal joint profit is 6 (when all players are cooperative) and decreases by one unit per aggressive player. Each player switching from C to A gains one additional unit of profit and decreases by one the profit of the other two.

Prove that the dominating strategy equilibrium is Pareto dominated and that the game has no strong equilibrium. Prove that there are exactly four imputations and that the α core coincides with the set of imputations.

Solution:

Strategy A strictly dominates C, so the unique Nash
equilibrium is (A, A, A). It is also the dominating strategy
equilibrium. However (A, A, A) is Pareto inferior to (C, C, C).
Thus, there is no strong equilibrium.

The guaranteed utility level is 1 (by using A), hence the
imputations

(C, C, C) (A, C, C) (C, A, C) (C, C, A)

(C, C, C) is in the α core. If Player 1 threatens to switch to
A, the counterthreat by {2, 3} is (A, A). No other coalition
has a profitable deviation.

(A, C, C) is in the α core. Again, coatitions with two
players cannot jointly change their strategy and enjoy a
benefit (each member being at least as well as before). To
a threat by Player 2 the counterthreat by {1, 3} is (A, A).

3) A symmetrical three-player game

Each player must pick one among the three players, possibly
himself. Hence $X_1 = X_2 = X_3 = \{1, 2, 3\}$. The payoffs of
the game are deduced from those listed below by the symmetrical
character of our game.

$$(x_1, x_2, x_3) = (1, 2, 3) \text{ payoffs: } (0, 0, 0)$$
$$= (1, 2, 1) \text{ payoffs: } (0, 0, -1)$$
$$= (1, 3, 1) \text{ payoffs: } (0, 2, 0)$$
$$= (1, 1, 1) \text{ payoffs: } (3, 1, 1)$$
$$= (1, 3, 2) \text{ payoffs: } (0, 2, 2)$$
$$= (2, 3, 1) \text{ payoffs: } (2, 2, 2)$$
$$= (2, 3, 2) \text{ payoffs: } (-1, 3, 3)$$

a) Prove that none of the Nash equilibria is Pareto-optimal. Thus the game has no strong equilibrium. b) Describe the imputations (five outcomes) and the α core (two outcomes).

Solution:

   (a)   The Nash equilibria are

      (1, 2, 3);   (1, 3, 2);   (3, 2, 1);   (2, 1, 3)

all are Pareto dominated by (2, 3, 1) or (3, 1, 2).

   (b)   The five imputations are (1, 1, 1)(2, 2, 2)(3, 3, 3). The Player's vote is unanimous: (2, 3, 1)(3, 1, 2). Each player receives exactly one vote. The α core contains (2, 3, 1) and (3, 1, 2).

b)   *Repeated games*

4)   <u>Repetition of the prisoner's dilemma</u>

   In the prisoner's dilemma (Example 2, Chapter 3) repeated over time, we assume that each player uses a stationary strategy with length 1 memory. Thus, Player i's strategy is a triple $(x_i; y_i, z_i)$ where x_i, y_i, z_i all belong to A, P , to be interpreted as follows. Player i plays $x_i = x_i^1$ in the first occurrence of the game (time t = 1). At time $t \geq 2$, he plays y_i if Player j was peaceful at time t - 1, and he plays z_i if Player i was aggressive.

$$x_i^t = y_i \quad \text{if } x_j^{t-1} = P$$

$$x_i^t = z_i \quad \text{if } x_j^{t-1} = A$$

For instance "tit for tat" is (P; P, A). The overall payoff is the Cesaro mean limit of instantaneous payoffs. a) Retain one copy of each pair of equivalent strategies and write the remaining 6 x 6 bimatrix game. b) Prove this game is dominance solvable and find its sophisticated equilibrium.

Solution:

(a) Strategy (P; A, A) is equivalent to (A; A, A). Also, (A; P, P) is equivalent to (P; P, P).

The reduced 6 x 6 matrix is

	PPP	AAA	PPA	*APA	*PAP	AAP	
PPP	2 / 2	0 / 3	2 / 2	2 / 2	0 / 3	0 / 3	
AAA	3 / 0	1 / 1	1 / 1	1 / 1	3 / 0	3 / 0	
PPA	2 / 2	1 / 1	2 / 2	a / a	a / a	a / a	a = 3/2
APA	2 / 2	1 / 1	a / a	1 / 1	a / a	a / a	*
PAP	3 / 0	0 / 3	a / a	a / a	a / a	0 / 3	*
AAP	3 / 0	0 / 3	a / a	a / a	3 / 0	a / a	

(b)   PAP is dominated by AAP; APA is dominated by PPA.
Hence the reduced form (where we omit Player 2's payoff,
which is found by symmetry),

PPP	2	0	2	0	*
AAA	3	1	1	3	
PPA	2	1	2	a	
AAP	3	0	a	a	

Here PPP is dominated by PPA, hence the new reduction

AAA	1	1	3	
PPA	1	2	a	
AAP	0	a	a	*

Now AAP is dominated by PPA, hence,

AAA	1	1
PPA	1	2

where PPA is the dominating strategy.

The sophisticated equilibrium is (PPA)(PPA)(Namely, tit-for-tat) with payoffs (3, 3).

c) β *core and s core in two-person games*

5) <u>A two-person game with quadratic payoffs</u>

A two-person game $X_1 = X_2 = R$

$$u_1(x_1, x_2) = - \{x_1^2 + 2ax_1x_2 + x_2^2 - 2x_1 - 2ax_2\}$$

$$u_2(x_1, x_2) = - \{x_1^2 + 2ax_1x_2 + x_2^2 - 2ax_1 - 2x_2\}$$

where a is a real parameter such that $|a| < 1$. a) Compute the Pareto-optimal outcomes and their associated utility vectors. b) Compute each player's optimal utility vector as a leader. Is the s core empty? If not, compute it.

Solution:

(a) As $|a| < 1$, u_i is a concave function of (x_1, x_2) for i = 1, 2. Thus a Pareto-optimal outcome maximizes

$(\lambda u_1 + (1 - \lambda)u_2)$ for some λ, $0 \le \lambda \le 1$ (and conversely any such outcome is Pareto-optimal). Since $\lambda u_1 + (1 - \lambda)u_2 = -(x_1^2 + 2ax_1x_2 + x_2^2) + 2(\lambda + a(1 - \lambda)x_1) + 2(a\lambda + (1 - \lambda))x_2$ is concave also, we need only the first-order conditions:

$$\left. \begin{array}{l} \lambda + a(1 - \lambda) = x_1 + ax_2 \\[2mm] a\lambda + (1 - \lambda) = ax_1 + x_2 \end{array} \right\} \quad <=> \quad x_1 = \lambda, \; x_2 = 1 - \lambda$$

Thus the Pareto set $x_1 + x_2 = 1$, x_1, $x_2 \ge 0$ with associated utilities

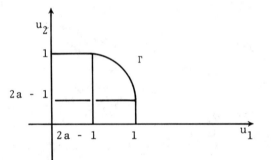

Γ is a parabola parametrized as

$$u_1 = -2(1 - a)\lambda^2 + 4(1 - a)\lambda + 2a - 1$$

$$u_2 = -2(1 - a)\lambda^2 + 1$$

Its equation is $8(u_2 - 1) = (u_1 - u_2 + 2(1 - a))^2$.

    (b)   Player 2's best reply

$$x_2 = r_2(x_1) \quad <=> \quad \frac{\partial u_2}{\partial x_2} = 0 \quad <=> \quad x_2 = 1 - ax_1$$

Hence Player 1 as a leader solves

$$\max_{x_1} u_1(x_1, 1 - ax_1) = -\{x_1^2 + 2ax_1(1 - ax_1) +$$

$$(1 - ax_1)^2 - 2x_1 - 2a(1 - ax_1)\}$$

$$= -\{(1 - a^2)x_1^2 - 2(1 - a^2)x_1 +$$

$$(1 - 2a)\}$$

His optimal first move is $x_1^* = 1$ so that $S_1 = 2a - a^2$. Similarly, $x_1 = r_1(x_2) \iff x_1 = 1 - ax_2$. Player 2 as a leader solves

$$\max_{x_2} u_2(1 - ax_2, x_2) = -\{(1 - ax_2)^2 + 2ax_2(1 - ax_2) +$$

$$x_2^2 - 2a(1 - ax_2) - 2x_2\}$$

$$= -\{(1 - a^2)x_2^2 - 2(1 - a^2)x_2 +$$

$$(1 - 2a)\}$$

$$\Rightarrow x_2^* = 1 \text{ and } S_2 = 2a - a^2$$

To check if (S_1, S_2) is feasible, search the intersection of Γ with $u_1 = u_2$. This gives $u_2 = 1 + \frac{1}{2}(1 - a)^2$. As $S_1 = 2a - a^2 < 1 + \frac{1}{2}(1 - a)^2$ always hold, there is no competition for the first move.

Note finally that $u_2 = S_2$ corresponds on Γ to

$$-2(1 - a)\lambda^2 + 1 = 2a - a^2 \iff \lambda = \sqrt{\frac{1 - a}{2}}$$

Thus the s core is the segment between $(x_1, x_2) = [\sqrt{(1 - a)/2}$,
$1 - \sqrt{(1 - a)/2}]$ and $[1 - \sqrt{(1 - a)/2}, \sqrt{(1 - a)/2}]$.

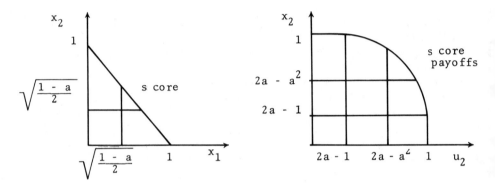

Case figure when $a > \dfrac{1}{2}$

In Exercises 6 and 7, G is a finite two-person game whose
utility functions are one to one on X_{12}.

6) More on the classification of two-person games

a) If G is in Class ii) prove

$$\beta_i < S_i \quad i = 1, 2$$

b) If each player has a (strictly) dominating strategy, then G
is in Class iii). c) If (S_1, S_2) is a Pareto-optimal payoff
vector, then G is in Class iii).

Solution:

(a) Suppose $\beta_2 \geq S_2$. Pick a 1-Stackelberg equilibrium

(x_1, x_2). Since x_2 is the best reply to x_1, $u_2(x_1, x_2) \geq \beta_2$. However, $u_1(x_1, x_2) = S$, so $(u_1, u_2)(x_1, x_2) \geq (S_1, S_2)$, and the competition for the first move does not arise.

(b) Let x_i^* be Player i's dominating strategy. Compute

$$\beta_i = \inf_{x_j} \sup_{x_i} u_i(x_i, x_j) = \inf_{x_j} u_i(x_i^*, x_j) = \alpha_i$$

Thus the competition for the second move does not arise. Also we have

$$S_i = \sup_{x_i} u_i(x_i, r_j(x_i)) = \sup_{x_i} u_i(x_i, x_j^*) = u_i(x_i^*, x_j^*)$$

Thus $(S_1, S_2) = (u_1, u_2)(x_1^*, x_2^*)$, and the competition for the first move does not arise either.

(c) We must prove (β_1, β_2) is a feasible payoff vector. If x is a 1-Stackelberg equilibrium, we have (just as in question a above)

$$(u_1, u_2)(x) = (S_1, u_2(x)) \geq (S_1, \beta_2)$$

Since (S_1, S_2) is Pareto-optimal, we must have $u_2(x) \leq S_2$. Thus $\beta_2 \leq S_2$ and a symmetrical argument shows $\beta_1 \leq S_1$.

7) Guaranteed deterring scenarios

Let (x^*, ξ_1, ξ_2) be a deterring scenario of G. We say that it is guaranteed if no player can suffer a loss by carrying out his threat:

$$u_1(x_1, \xi_2(x_1)) \leq u_1(x^*) \leq u_1(\xi_1(x_2), x_2)$$

for all x_1, x_2

$$u_2(\xi_1(x_2), x_2) \leq u_2(x^*) \leq u_2(x_1, \xi_2(x_1))$$

The g core of G is the subset, denoted C_g, of those outcomes x^* in the α core such that there exists at least one guaranteed deterring scenario (x^*, ξ_1, ξ_2). a) Prove the following equivalence $x \in C_g \iff [x$ is Pareto-optimal, and $u_i(x) \leq \beta_i]$. b) Prove that the g core is a — possibly empty — subset of the s core:

$$C_g \subset C_s$$

c) If G is in Class i), prove by examples that the g core can be empty or nonempty. d) If G is in Class iii), prove that either its g core is empty, or $\beta = (\beta_1, \beta_2)$ is the payoff vector of at least one Pareto-optimal outcome. In the latter case, prove that $C_g = C_s = C_\beta = (u_1, u_2)^{-1}(\beta)$.

Solution:

(a) If $x^* \in C_g$, then it is Pareto-optimal and moreover for all x_2

$$u_1(x^*) \leq u_1(\xi_1(x_2), x_2) \leq \sup_{x_1} u_1(x_1, x_2)$$

Thus $u_1(x^*) \leq \beta$. The proof of $u_2(x^*) \leq \beta_2$ is identical. Conversely, say that x^* is Pareto-optimal and satisfies $u_i(x^*) \leq \beta_i$, $i = 1, 2$. Then one can construct a pair of threats ζ_1, ζ_2 such that

$$u_1(x^*) \leq u_1(\zeta_1(x_2), x_2) \qquad \text{all } x_2$$

$$u_2(x^*) \leq u_2(x_1, \zeta_2(x_1)) \qquad \text{all } x_1$$

The remaining two inequalities follow by the Pareto-optimality of x^*.

(b) Let $x^* \in C_g$, and x be a 1-Stackleberg equilibrium. We have

$$u_2(x) \geq \beta_2 \geq u_2(x^*) \Rightarrow u_1(x^*) \geq u_1(x) = S_1$$

(the implication because x^* is Pareto-optimal). The proof of $u_2(x^*) \geq S_2$ is identical.

(c) In a two-person zero-sum game without a value, any outcome x such that $\alpha_1 \leq u(x) \leq \alpha_2$ is in the g core.

Here is an example of a game in Class i (the β core is empty) where the g core is empty as well.

0		2	
	2		0
3		1	
	1		3

(d) We consider a game where the β core, s core and g core are all nonempty. We must prove that they all are equal to the singleton $(u_1, u_2)^{-1}(\beta)$.

Pick $x \in C_g$. We know $u_i(x) \le \beta_i$ $i = 1, 2$ and x is Pareto-optimal. Thus both inequalities are equalities (otherwise the β core is empty) and the g core, as well as the β core are equal to $\bar{x} = (u_1, u_2)^{-1}(\beta)$.

We show next that $\bar{x}$ is the i-Stackelberg equilibrium of G for $i = 1, 2$, which establishes $S_i = \beta_i$, $i = 1, 2$ and concludes the proof.

By finiteness of X and the one-to-one character of u_i, there is a unique strategy, namely $\bar{x}_2$ such that

$$\sup_{y_1} u_1(y_1, \bar{x}_2) \le \beta_1$$

Thus $u_1(\bar{x}_1, \bar{x}_2) = \beta_1 \ge u_1(y_1, \bar{x}_2)$ all y_1, and $(\bar{x}_1, \bar{x}_2)$ is on the best reply curve of Player 1. Every other outcome x on this best reply curve is such that $u_1(x) > \beta_1 \Rightarrow u_2(x) < \beta_2$. Thus $\bar{x}$ is Player 2's preferred outcome on this curve, so that $S_2 = \beta_2$, and we are home.